POCKET GUIDE TO FISHING LAKES & RESERVOIRS

Phoenix, Md.

Distributed exclusively to the book trade by Stackpole Books, 5067 Ritter Road, Mechanicsburg, PA 17055 1-800-732-3669

Special thanks to A.F.T.M.A. for their help in allowing us to use parts of their glossary and "Aquatic Education Curriculum". Also, thanks to Chris Duffy, an up and coming outdoor writer, and Mark Susinno, a young artist with great promise!

Cover Painting by
Mark Susinno

ILLUSTRATION AND PHOTO CREDITS

Berkley, Inc.-12,13,19; Bill Burton-5,26,32,37,62,70; Cary de Russy-6; Eagle Electronics-58; Fisher Marine-1; Louis Frisno-22,23,24; Jim Guilford-14; Dean Lee-5; Johnson Outboard Motors-80; Lowrance Electronics-5,66; Dave May-20; Mercury Marine-2; Rob Merz-25,28,29,30,31; O. Mustad & Son-9; PA Fish Comm.-25; Boyd Pfeiffer-48; Sampo-9; Somerville Studios-8,9; Mark Susinno-Cover, 28,34,35,38, 40,43,45,46,51,53,55,56,64,68,69,72,73,74,75,76,77,78,79; Zebco Corp.- 16,17,18,19

Published by FIM Publishing, Inc.
P.O. Box 197, Phoenix, MD 21131

Printed in the United States of America

Pocket Guide to Fishing Series
ISBN 0-917131-00-2

Lakes & Reservoirs
ISBN 0-917131-01-0

To Our Young Readers

If you're into drugs or alcohol, chances are this book won't do you much good. It's not that you won't learn something-you will. But, if your head isn't clear, you will miss the tranquility of a lake at sunrise, the serenity of a clear mountain stream, not to mention the delicate strike of a fish on the end of your line. It's really very simple, fishing provides its own natural high! No need for some mind altering drug. Besides, fishing under the influence of drugs or alcohol can be very dangerous.

Some people say that youngsters, or for that matter adults, take drugs to forget their problems. If you think this is the case, forget it! Your problems will still be there when you sober up, along with possibly a new one- drug dependency.

A lot of adults don't know how to fish. If you have never been, chances are the people raising you fall into this category. Why don't you suggest to them that you all learn together. Funny thing about fishing-with a little instruction, which this book will provide, anyone can be successful. You learn as you do it. You also get away from this hectic world we live in and get a chance to really relax. It's a great way to get to know someone, like the people who are trying to raise you.

Marcel C. Malfregeot Jr.
Administrative Assistant
Harrison County Schools
Clarksburg, WV

CONTENTS

1

SELECTING TACKLE

Shopping for fishing tackle can be almost as much fun as fishing. Manufacturers are turning out equipment today that is stronger, lighter and more sensitive than at any time in history. That is a long time, since history suggests that rods and reels were in use in China since the 3rd or 4th century A.D.

Your tackle shop can show you equipment that was designed with the aid of a computer and contains materials that were developed as a result of the space program. Always buy the best equipment that you can afford. If you follow the manufacturer's instructions for taking care of it, you may be assured of many years of faithful service.

It is very important that your rod, reel and line match each other. This is called **BALANCED TACKLE.** To understand this concept, let's look at a practical example. The rod you have selected is rated by the manufacturer for 6, 8 and 10 pound test line. The reel capacity shows the spool will handle 6, 8, 10 and 12 pound line. Being able to use line of different breaking strengths will give you the flexibility to match your equipment to the many different conditions you might encounter.

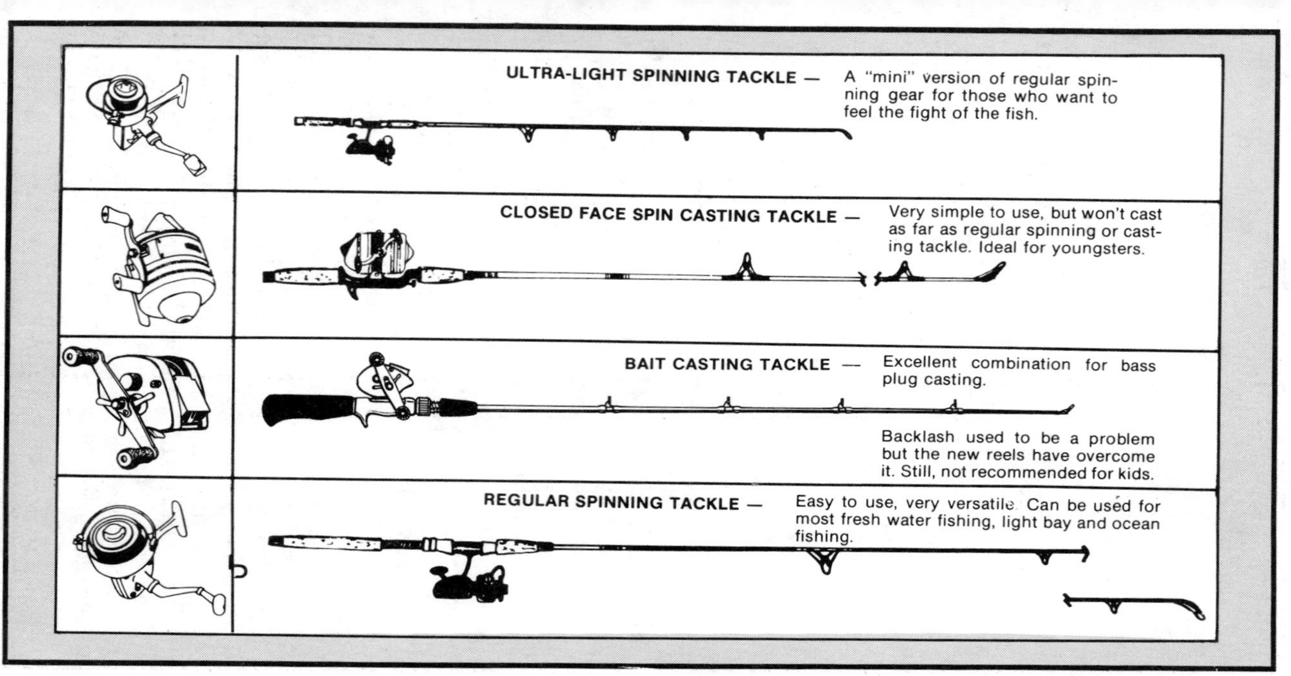
ULTRA-LIGHT SPINNING TACKLE — A "mini" version of regular spinning gear for those who want to feel the fight of the fish.
CLOSED FACE SPIN CASTING TACKLE — Very simple to use, but won't cast as far as regular spinning or casting tackle. Ideal for youngsters.
BAIT CASTING TACKLE — Excellent combination for bass plug casting.
Backlash used to be a problem but the new reels have overcome it. Still, not recommended for kids.
REGULAR SPINNING TACKLE — Easy to use, very versatile. Can be used for most fresh water fishing, light bay and ocean fishing.

The drag setting on the push button reel is on top or near the handle. On the open-face reel, you will find it in front or in the rear.

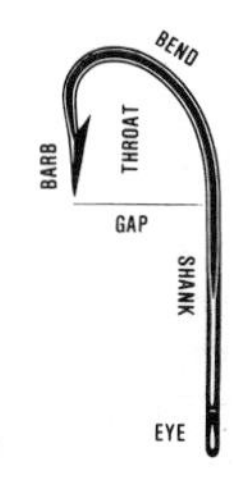

Anatomy of a fish hook.

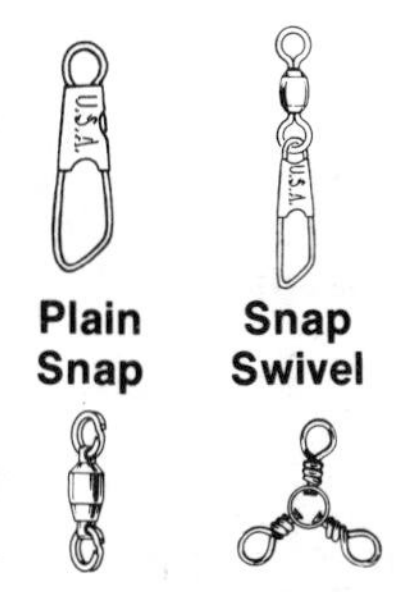

Plain Snap

Snap Swivel

Barrel Swivel

3-Way Swivel

Spin fishing is the most popular type fishing in the United States, and the simplist. Because you are just beginning, we will confine our attention to this type of equipment.

There are two basic kinds of spinning reels. The closed-face or push button reel is by far the easiest to use. The open-faced reel is preferred by most experienced anglers because it casts further .

Rods come in all sizes. A typical rod for a beginner should be between 5½ to 7 feet with a medium to light-medium action. When choosing a reel, try to select one that will hold at least 200 yards of line.

The most popular fishing line in use today is monofilament line. It is made from nylon. Always buy the best you can afford and check it often for knots and scrapes, usually in the first few feet. Fishing line is graded in different **pound test** categories. If the line is 6 pound test line, ideally it should break if you pick up a weight that weighs more than 6 pounds.

Sinkers are lead weights used to get your hook down to the proper depth and put enough tension on your line so you can feel the fish strike.

Swivels are used to prevent your fishing line from twisting. They come in bright plated and dull brown or black finishes. Some fishermen think the bright finishes add to their offering, others don't.

All of these items come in different sizes

There are over 50,000 different sizes and styles of hooks to choose from. The most important factor in selecting a hook is its size.

There are two different numbering systems for hooks. Smaller hooks start with the #1 and size down in even numbers to #28, the smallest. Larger hooks have a /0 after the number. The bigger the number, the bigger the hook is.

When in doubt, always select a hook the next size smaller because larger fish can be caught on small hooks, but smaller fish can't swallow a large hook. They can, however, nibble away all of your bait.

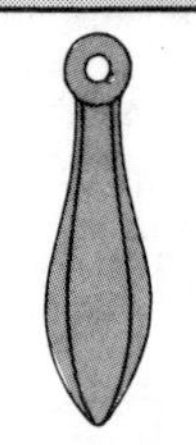

Bank Sinker: An all around favorite. Holds bottom well.

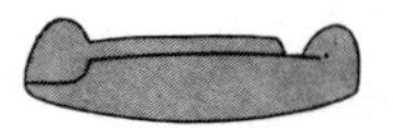

Clinch on: Soft lead ears are clamped on line for casting.

Rubber Core Sinker: Preferred by many because it can be slipped on and off line easily.

Split Shot: Designed to be clinched on for casting.

Bobbers hold your bait at a desired depth.

SHOPPING LIST

BALANCED ROD, REEL AND FISHING LINE

LAKES & RIVERS
5½' - 7' rod with 6 lb. to 14 lb. line

SALT WATER
6' - 7' rod with 14 lb. to 20 lb. line

TACKLE BOX Big Enough for Additional Tackle

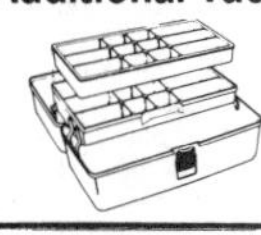
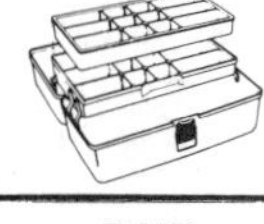

LAKES & RIVERS
SALT WATER
Make sure it is worm proof.

BAIT BUCKET

Useful in fresh & salt water. Keeps live bait, like minnows alive.

SWIVELS Packages of Each

LAKES & RIVERS
Sizes #12 - #10 Plain Snap for lures; Snap Swivel for bottom rigs.

SALT WATER
Sizes #7, #1, 1/0 snap swivels for bottom rigs.

SINKERS Assorted Packs

LAKES & RIVERS
Split Shot or Pinch On ¼ oz. - ¾ oz.
Bank Sinker Kind ⅛ oz. - ¾ oz.

SALT WATER
Split Shot or Pinch On ½ oz. - 1 oz.
Bank & Pyramid 1 oz. - 4 oz.

HOOKS Pack of Assorted Pack of Snelled

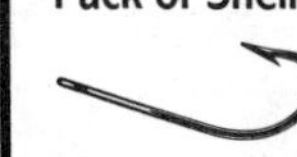

LAKES & RIVERS
Pack of plain hooks #8 - #2, 2/0
Pack of Snelled for bottom rig: Sizes #8 - #2, 2/0

SALT WATER
Snelled, Sizes 1/0 - 5/0 with wire leader.

LINE CLIPPERS

Always good to have for snipping off excess line.

GOOD KNIFE

Good for cutting live bait and filleting catch.

BOBBERS 3 Different Sizes

Normally used with bottom rigs. Match up with sinker sizes.

BOTTOM RIGS (3)

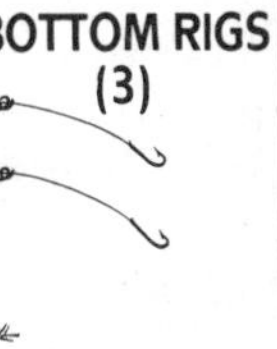

LAKES & RIVERS
Usually called Crappie Rigs. Normally comes with hooks.

SALT WATER
Usually called Top & Bottom Rigs. Should have wire leader.

NEEDLE NOSE PLIERS

Has many uses. Good for removing hooks and lures from fish's mouth.

STRINGER OR BUCKET

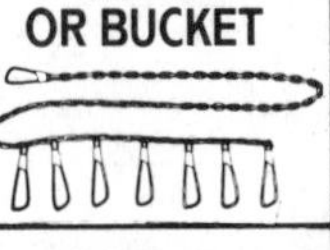

LAKES & RIVERS
Either

SALT WATER
Bucket

ARTIFICIAL LURES

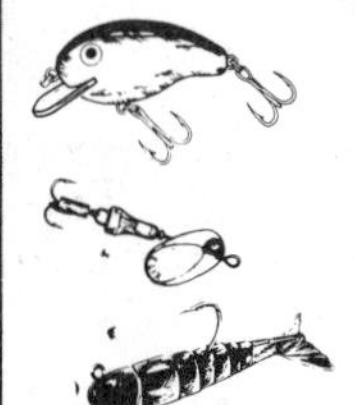

LAKES & RIVERS
1) Single Shaft Spinner
2) Plastic Worms. Sizes 6"-8" with 1/0-4/0 Worm Hooks and ⅛ - ⅝ Slip Sinkers.
3) (1 each) ¼-⅜ oz. Surface Lure. Medium Diver & Deep Diver.

SALT WATER
Mainly used in trolling from boat.

2
KNOTS

The knot is the weakest point in your fishing line and one of the main reasons for fish getting away. It makes sense that you should tie the strongest most reliable knot you can.

The knots demonstrated on this page and the next should cover most situations you might encounter. In the next chapter we will go into the proper casting techniques. It will be suggested to you that casting requires a lot of practice that should take place before you go fishing. It is also a good idea to become as familiar as you can with the knots that can make or break a fishing trip.

Improved Blood Knot

The Improved Blood Knot is used for tying two pieces of monofilament together of relatively equal diameters.

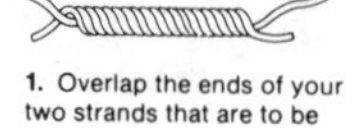

1. Overlap the ends of your two strands that are to be joined and twist them together about 10 turns.

2. Separate one of the center twists and thrust the two ends through the space as illustrated.

3. Pull knot together and trim off the short ends.

Arbor Knot

The Arbor Knot provides the angler with a quick, easy connection for attaching line to the reel spool.

1. Pass line around reel arbor.

2. Tie an overhand knot around the standing line.

3. Tie a second overhand knot in the tag end.

4. Pull tight and snip off excess. Snug down first overhand knot on the reel arbor.

Trilene® Knot

The Trilene Knot is a strong, reliable connection that resists slippage and premature failures. It works best when used with Trilene premium monofilament fishing line.

The Trilene Knot is an all-purpose connection to be used in joining Trilene to swivels, snaps, hooks and artificial lures. The knot's unique design and ease of tying yield consistently strong, dependable connections while retaining 85-90% of the original line strength. The double wrap of mono through the eyelet provides a protective cushion for added safety.

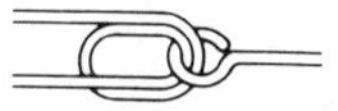

1. Run end of line through eye of hook or lure and double back through the eye a second time.

2. Loop around standing part of line 5 or 6 times.

3. Thread tag end back between the eye and the coils as shown.

4. Pull up tight and trim tag end.

Palomar Knot

The Palomar Knot is a general-purpose connection used in joining monofilament to swivels, snaps, hooks and artificial lures. The double wrap of mono through the eyelet provides a protective cushion for added safety.

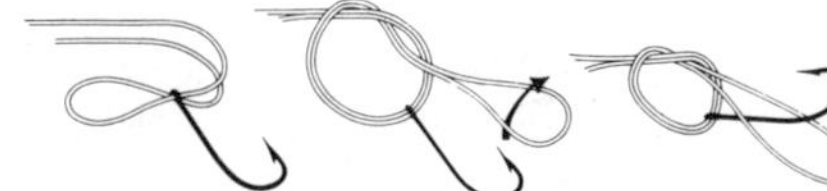

1. Double the line and form a loop three to four inches long. Pass the end of the loop through hook's eye.

2. Holding standing line between thumb and finger, grasp loop with free hand and form a simple overhand knot.

3. Pass hook through loop and draw line while guiding loop over top of eyelet.

4. Pull tag end of line to tighten knot snugly and trim tag end to about ⅛".

Double Surgeon's Loop

The Double Surgeon's Loop is a quick, easy way to tie a loop in the end of a leader. It is often used as part of a leader system because it is relatively strong.

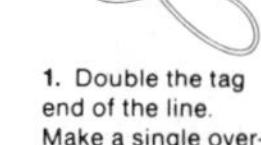

1. Double the tag end of the line. Make a single overhand knot in the double line.

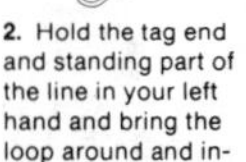

2. Hold the tag end and standing part of the line in your left hand and bring the loop around and insert through the overhand knot again.

3. Hold the loop in your right hand. Hold the tag end and standing line in your left hand. Moisten the knot (don't use saliva) and pull to tighten.

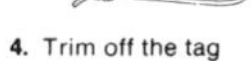

4. Trim off the tag end.

3
HOW TO USE YOUR ROD & REEL

It may seem pretty silly to go home with your new rod and reel and spend the afternoon casting on your lawn or sidewalk. Don't let that bother you. You would feel a lot sillier if the first cast you made when you went fishing got hung up-halfway up a 40 foot tree. This sometimes happens to the most experienced angler. But obviously, the more you practice, the less chance you have of making a habit of decorating the landscape with your tackle.

As you become more experienced in reservoir fishing, you will discover that many species of fish, such as bass, like heavy cover. What this means is that you will find these fish in areas difficult to cast too. Thus, the angler who is an accurate caster has a better chance of catching fish than the one who isn't.

A lot of rod and reel combinations come with a practice casting plug. If your new outfit doesn't have one, it's a good idea to buy one. They don't cost much.

After you have mastered casting, try some target practice with a bath towel or garbage can lid and cast to it until you hit it consistently. Now, try moving closer and then further away until you hit it consistently from these different distances.

Switch to a smaller target. You will be surprised at how rapidly you will learn to drop the plug just where you want it.

Earlier you were shown the drag setting on both the opened faced and push button reels. The drag is a mechanical device on your reel that can be adjusted to control the resistance to the turning of the reel spool. It allows for the reel to let line out rather than break when fighting a fish. A good way for a beginner to judge the proper drag setting is to tighten it so that when you try to yank the line out of the reel, it breaks. Now, loosen the drag until it is just loose enough so that when you yank the line, it doesn't break. Proper drag settings will change with different test lines.

Playing and landing a fish takes practice. Most fishermen use the **pump and reel** technique. After the fish is played out, lower your rod tip and reel in at the same time. Now, without allowing any slack in the line, begin the process again and continue it until the fish is landed.

OVERHEAD CAST WITH CLOSED FACE SPINNING REEL

For the more accurate two-handed cast, hold your rod and reel as shown with reel handles pointing up and depress push button with thumb. With other hand, take line lightly between thumb and index finger.

With a slightly angled body, lift the rod until the tip is just above target (10 o'clock). Your elbow and upper arm should be close to your body, and your forearm parallel to angle of the rod.

3

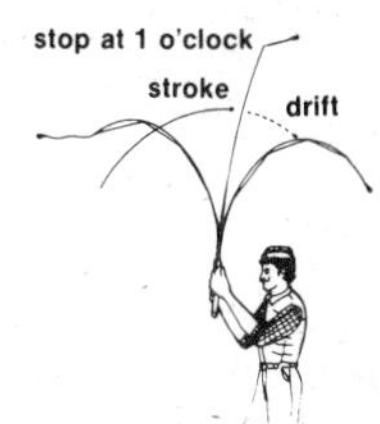

Lift your arms smoothly until your hands are at eye level. Stop the rod at 1 o'clock and allow the momentum of your bait to flex the rod tip backwards.

4

Without hesitation, stroke forward quickly and release push button at 11 o'clock to set your bait in flight.

5

Follow through by lowering the rod tip to follow the flight of the bait. If the bait goes straight up, you released too soon; if it flops in front of you, you released too late.

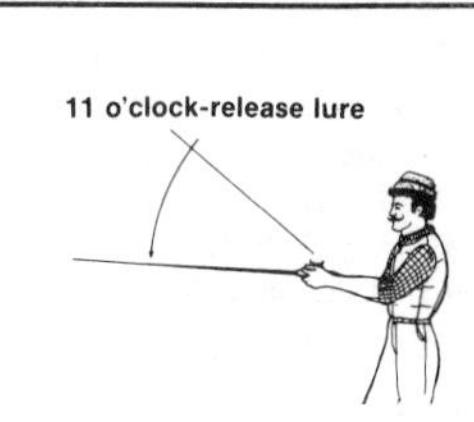

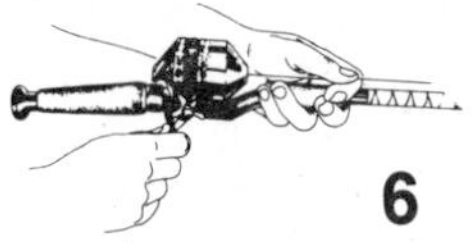

6

As your bait nears the target, apply pressure to the line with the thumb and index finger of your other hand for accuracy.

7

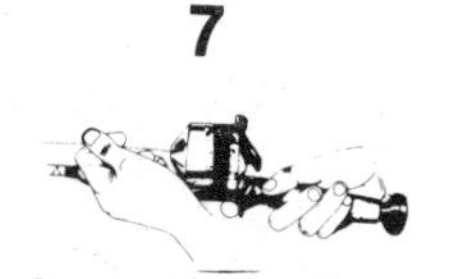

As you begin retrieve, let line flow through thumb and index finger of your other hand.

Pretend you are standing on 6 o'clock with your head under 12 o'clock. Your rod should be pointed at 10 o'clock. Bring it back to 1 o'clock, bring it forward and release at 11 o'clock.

OVERHEAD CAST WITH OPEN FACE SPINNING REEL

1

Grip the rod and reel as shown. With free hand rotate reel's cowling until line roller is beneath extended index finger. Pick up line with that finger and flop open bail with other hand.

2

Stand with body angled slightly toward target. Center the rod with tip top at eye level (10 o'clock). Position elbow close to your side; your forearm in line with the rod.

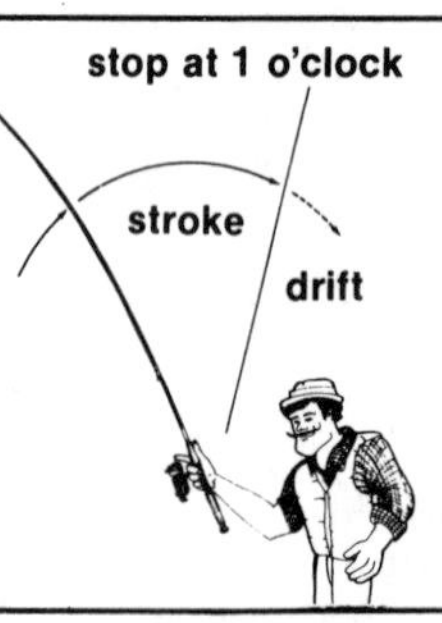

3

Begin by swiftly raising head almost to eye level, pivoting elbow.

4

When the rod reaches 1 o'clock, the weight of the bait will cause it to bend to rear. At this time bring rod forward in a crisp downstroke.

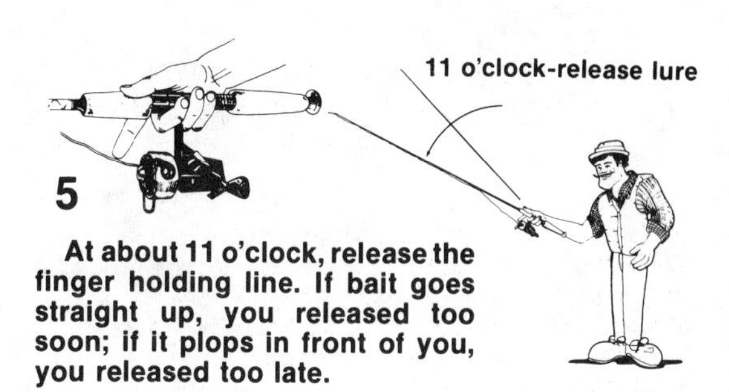

At about 11 o'clock, release the finger holding line. If bait goes straight up, you released too soon; if it plops in front of you, you released too late.

As your bait nears target, gently "feather" line with index finger. The moment it hits target, place index finger on edge of spool to stop flight of bait and prevent slack build-up on reel spool.

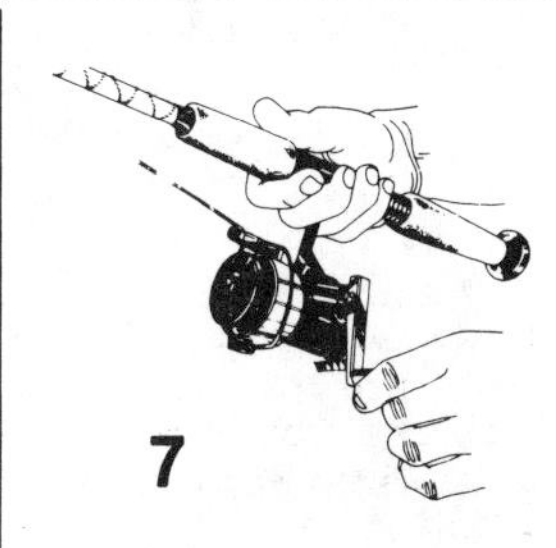

Without changing hands, begin retrieve. The line guide will automatically flop over.

PLAYING AND LANDING A FISH

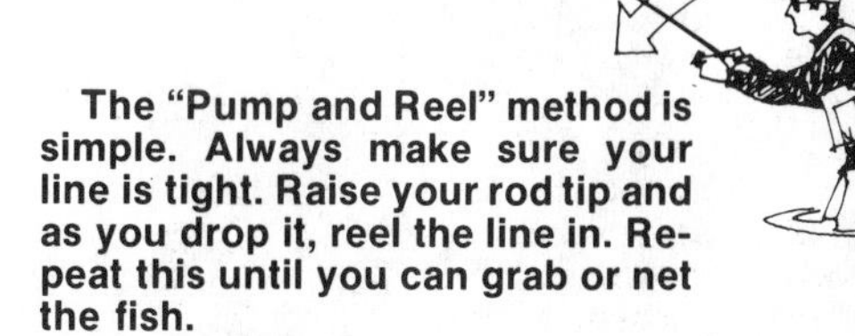

The "Pump and Reel" method is simple. Always make sure your line is tight. Raise your rod tip and as you drop it, reel the line in. Repeat this until you can grab or net the fish.

4

FISH IDENTIFICATION

Imagine being out on the water and hooking into a whopper. You fight your quarry for what seems to be hours and finally land him. You go back to the dock with a big grin on your face only to have it removed when someone asks you what kind of fish it is and you don't know. What do you say?

It can be embarrassing if you can't identify the different species of fishes you might catch and it happens all the time, even to some experienced anglers. One reason for this is that within a family of fish there may only be slight differences in color or features.

There are many different species and subspecies of fish in the lakes, ponds and reservoirs of this country. It would be impossible to cover them all in this short chapter. On the following pages are illustrations of some of the more common species found in the United States.

Be sure to study these illustrations and the notations on each individual fish carefully. This is important because in some cases the major difference between two fish from the same family can be the number of spines on the dorsal fin.

It would also be a good idea to spend some time studying the interior and exterior anatomy of fish found in the illustrations at the end of this chapter.

LARGEMOUTH BASS

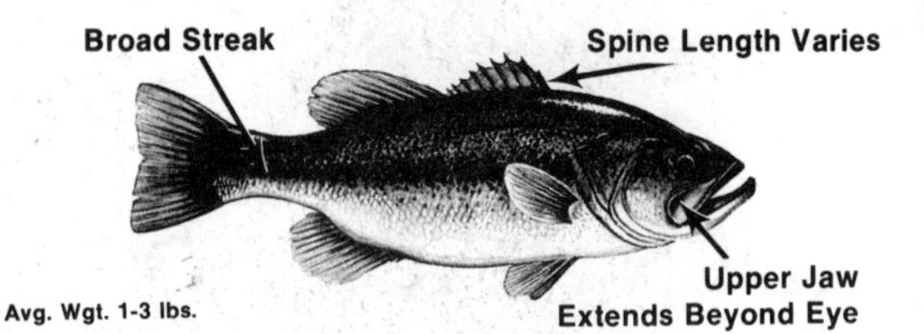

Avg. Wgt. 1-3 lbs.

SMALLMOUTH BASS

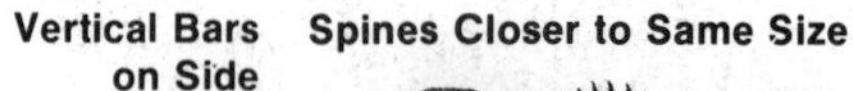

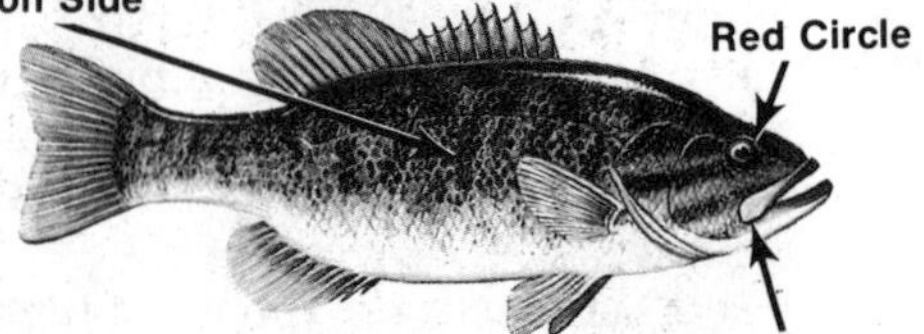

Avg. Wgt. 1-2 lbs.

BLACK CRAPPIE

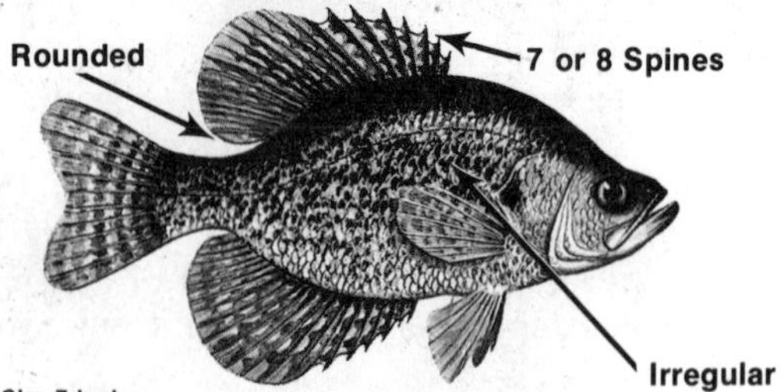

Avg. Size 7 inches

YELLOW PERCH

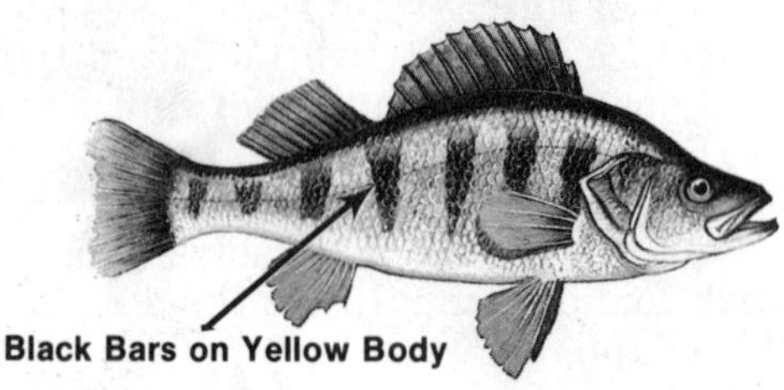

Avg. Size 8 inches

BLUEGILL

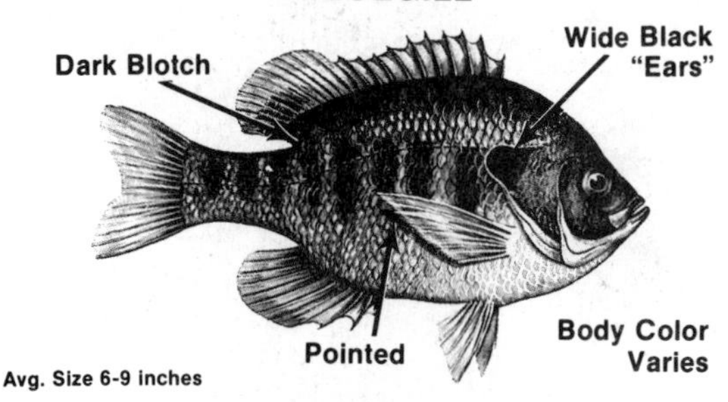

Avg. Size 6-9 inches

PUMPKIN SEED

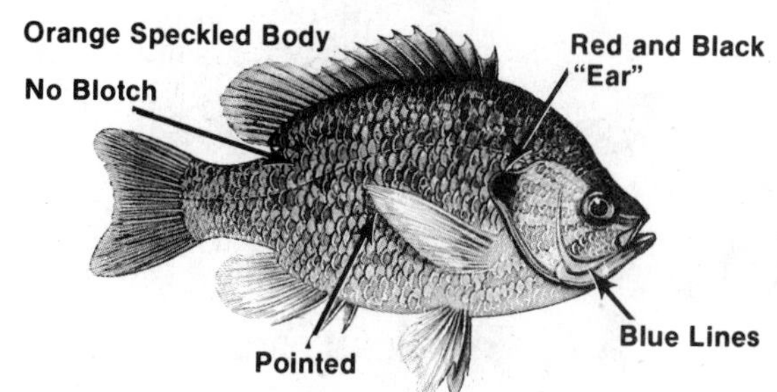

Avg. Size 6-7 inches

STRIPED BASS

Avg. Wgt. 2-5 lbs.

CHANNEL CATFISH

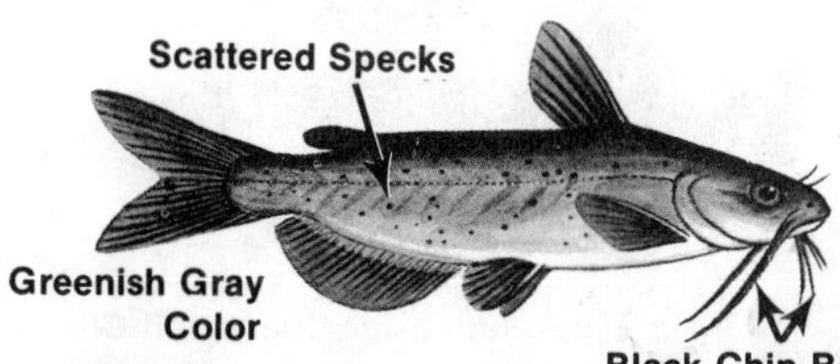

Avg. Wgt. 1-3 lbs.

CARP

Avg. Wt. 5-8 lb.

WALLEYE

Avg. Wt. 3 lb.

NORTHERN PIKE

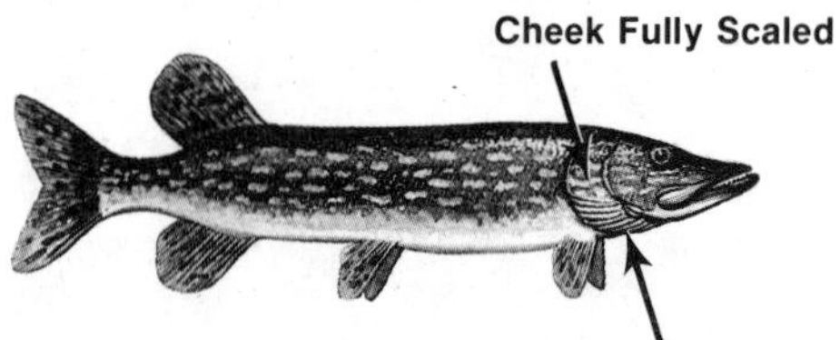

Avg. Wgt. 5 lb.

MUSKELLUNGE

Avg. Wgt. 8 lb.

EXTERIOR FEATURES

SOFT-RAYED FISH

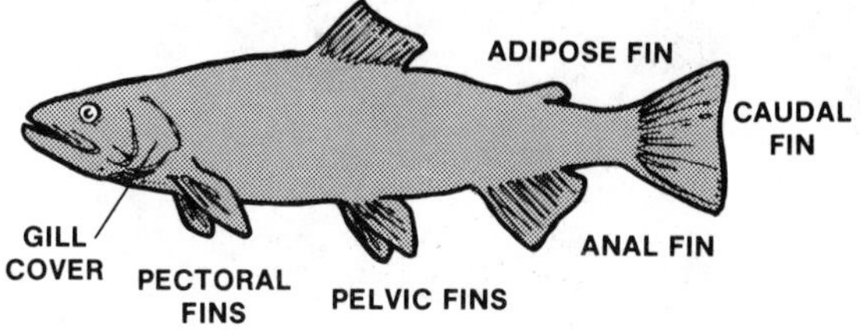

SPINY-RAYED FISH

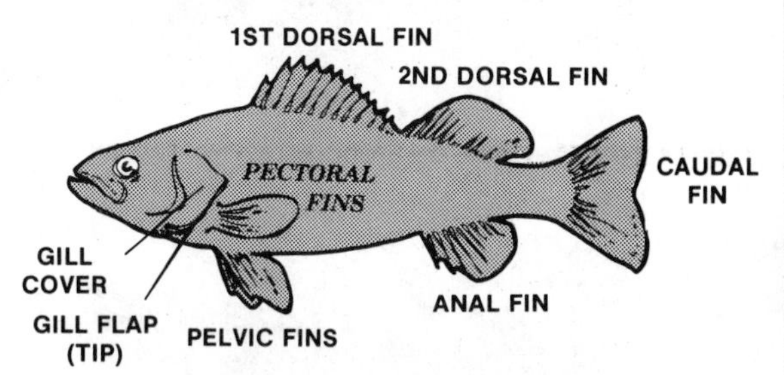

INTERIOR FEATURES

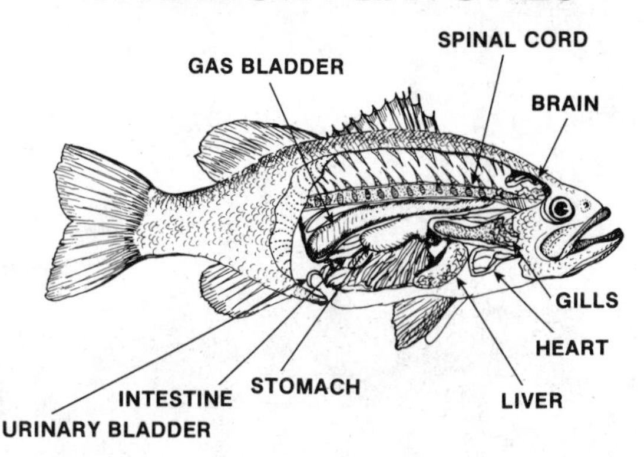

5

LIVE BAIT FACTS

Live bait is the most effective lure you can use. Unlike artificial baits, which attempt to imitate natural food, live bait represents the food fish normally seek.

No bait will automatically catch fish. The observant fisherman will try to discover his quarry's preference and match the food on which the fish are presently feeding. This preference will change as the season progresses from **ice out** in early spring on the lakes, ponds and reservoirs in this country.

The **life cycle** of these waters has a definite pattern to it and it is important that you be aware of it. In most parts of the country, the primary source of early season food is minnows. As the weather warms, insects begin to hatch, crayfish start to move and the diet of fishes becomes more varied. As fall appraches, the cycle will reverse itself and once source of food after another will disappear.

Most fish like their groceries fresh. This involves taking good care of your bait from the moment you buy it or catch it yourself. All bait should be kept cool.

As important as keeping your bait fresh is the way you rig it. If it isn't rigged properly, it won't act properly when presented to the fish.

SHINERS

An excellent bass bait. The size of the shiner will determine the size of the hook used. With small ones use a 3/0 hook, with larger ones try a 5/0 hook.

If using a bobber, use the "top hook" technique. If casting a shiner, try the "lip hook" approach.

MINNOWS

Great for bass, perch, northern pike and trout, to name a few. Minnows are smaller and more delicate than shiners. The "lip hook" technique for casting is the same as a shiner, however, the "top hook" technique for bobber fishing differs. Because this bait is so delicate, it is best to hook it behind the dorsal fin.

CRAWFISH

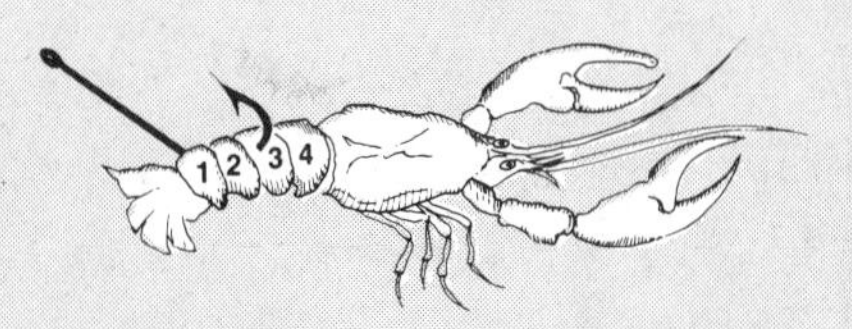

Great for bass and panfish. Reserve large ones for bass and smaller ones for whatever panfish is in your area. Rig them as shown in the illustration.

You can buy crawfish at most tackle shops or catch them in a nearby stream. You will find them under rocks during the day and with a flashlight at night when they roam around.

WORMS

Earth Worms, Night Crawlers

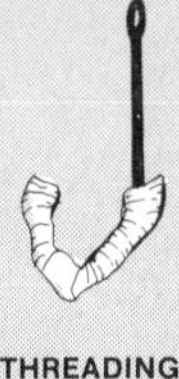

THREADING THE HOOK

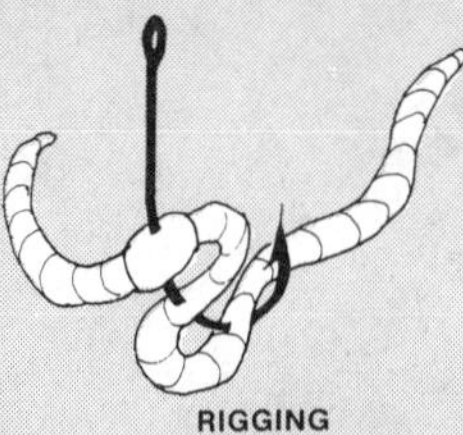

RIGGING FOR LARGER FISH

By far the most popular live bait used by fresh-water fishermen. There are several ways to correctly hook a worm. The more popular methods are illustrated above.

For small fish thread a section of the worm on your hook so they can't nibble away at your bait.

For larger fish use the method above, in the righthand illustration. For fish such as catfish and carp, fill the hook with several worms hooked using this method.

GRASSHOPPERS AND CRICKETS

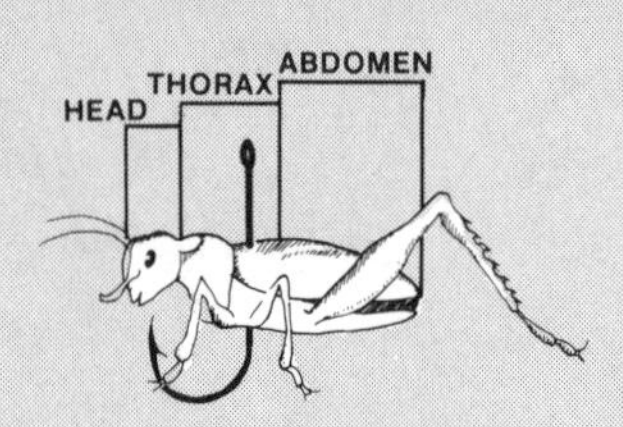

A lot of bait shops sell crickets and you can catch your own grasshoppers. They are very good baits and should be hooked through the thorax, as shown above. Make sure the hook barb is pointed forward.

When fishing a lake with these insects, add a split shot sinker to your line and fish it with a bobber.

HELLGRAMMITE

Watch out, they bite.

"COLLAR HOOK" FOR BOBBER FISHING

"COLLAR PIERCE" FOR CASTING

In many parts of the country hellgrammites are available. They are great bass bait.

For bobber fishing, hook the hellgrammite under its collar as shown above. If you plan to cast, pierce its collar as shown in the righthand illustration.

GRASS SHRIMP

SINGLE SHRIMP

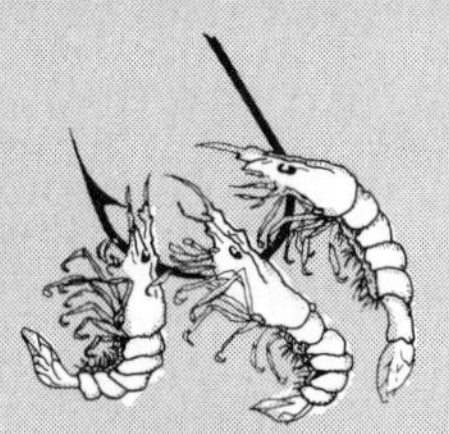
TRIPLE HOOK

Grass shrimp are very small and almost transparent when fresh. They are a great freshwater bait. If you are after panfish, hook a large one as shown in the lefthand illustration. If you are after larger fish, hook two or three grass shrimp on your hook as shown in the righthand illustration.

HOOK SELECTION

There is no fixed rule in hook selection and size. It's a good idea to use a heavier hook with heavy tackle and a lighter one with lighter tackle. Here is a general size chart to get you started.

HOOK SIZE	SPECIES
#4 - 1/0	Small bass
#3 - 4/0	Larger bass
#8 - #6	Smaller bluegill, perch & crappie
#4 - 1/0	Larger bluegill, perch & crappie
#6 - #2	Smaller catfish
2/0 - 4/0	Larger catfish
2/0 - 6/0	Muskie & northern pike
#6 - #2	Walleye
#12 - #4	Trout

6

FISHING RIGS

No matter how good, expensive, or expertly used a rod and reel is, the total fishing experience is only as good as the combination of line, sinker, bobber, hook and bait. Known as live bait rigs, these combinations of various elements of **terminal tackle** can be made by the angler or in some cases, purchased at your local tackle shop. These rigs can be cast, fished at various depths, with the aid of a bobber or float, or fished on the bottom of a reservoir or lake.

As you will learn later on, fish react to their environment. During the course of a year, lakes and reservoirs experience seasonal changes that will dictate the holding areas of game fish. As you will also learn later on, fish require food, oxygen and the proper temperature range. There will be a variance in need among species for other elements such as structure, pH and light levels. The rigs illustrated on the following pages are designed to allow you an adequate selection to meet most any need. In early spring, for instance, fish will be deep and moving about. At this time it is important to identify how deep they are. The casting rig, worked at various depths, is ideal for this.

Just as no two people do everything alike, you will probably also run into variations of these rigs, depending on the part of the country you live in. But, chances are, upon examination of the local interpretation, you will probably find little real difference.

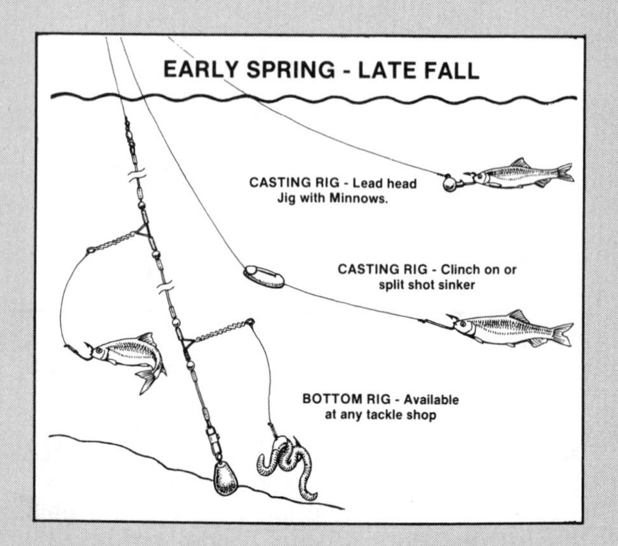

EARLY SPRING - During this period of the year, as you will learn, fish such as walleye, muskie and pike are spawning. Other fish, such as bass, are in deep water, moving about. Walleye will return to deep water during daylight hours because of their sensitive eyes.

Because the fish are deep and moving about, you want your bait to do the same. The best rig for this type of fishing is one of the above. Start off at the bottom. Make sure your sinker is heavy enough to get your bait that far down. Now, retrieve slowly counting the times the handle on your reel makes a full revolution. Make sure to stop every 2 or 3 revolutions. Do this until you get a strike. Chances are that you have found the depth at which the fish are suspended, and after you have landed the fish, you will know at what depth to find them again.

The casting rigs are the ones you will want to use for fishing the shallower spawning areas. The main difference is that your sinker will be much smaller. Remember to select the live bait that the fish are feeding on at the time. The bait dealer can tell you what it is. Again, a slow to moderate retrieve is recommended.

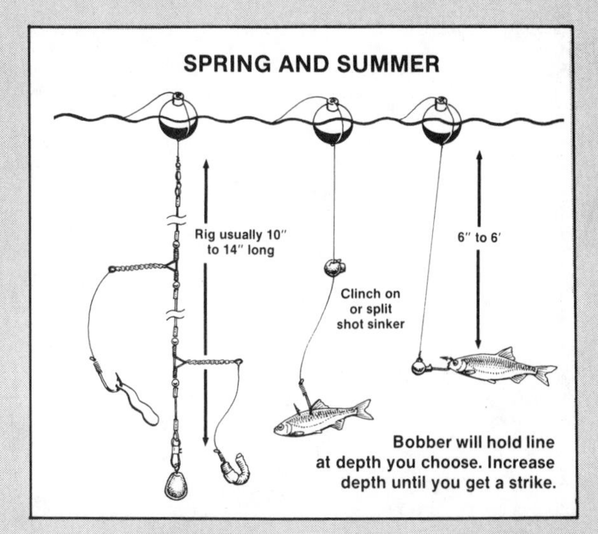

SPRING AND SUMMER - The rigs recommended for early spring are still the ones to use for spawning fish such as largemouth bass. But, as the water warms, fish will suspend at certain depths.

If they are close to the bottom, the bottom rig as shown above is a good choice. This rig is a small version of the bottom rig used in saltwater fishing. In some parts of the country it is called a "crappie" or "panfish" rig. Make sure you keep your line tight so that you can feel the fish taking your bait. Also, be sure your sinker is heavy enough to hold the bottom.

Bobbers are used to hold your bait or rig at a certain depth. They may be adjusted. If you find the fish are holding above the bottom, you may also use the bottom rig with a bobber. Now it's a matter of choosing a sinker that is just heavy enough to hold the bobber upright.

LATE FALL AND EARLY WINTER - During this period of the year fish will school up (gather) on points and bars. They will normally be in 15 to 25 feet of water and feeding actively. Use either of the rigs you used for fishing the spawning areas with a heavier clamp on or split shot sinker and retrieve your rig slowly.

7
ARTIFICIAL BAIT FACTS

Fishermen have been devising lures to imitate natural baits for centuries. In fact, the first fisherman to use artificial bait was probably someone who couldn't find a worm.

Artificial bait is fun to fish because you provide much of the action needed to make the lure act as if it were alive. This group includes jigs, spoons, crankbaits, plastic worms and tails and spinners. There are enough variations and combinations of these artificials to fill a book of their own.

They can be cast, trolled, jigged, popped, jerked, twitched, retrieved fast, slow, shallow, or somewhere inbetween. Many of these lures have some type of action built in by the manufacturer. However, they still leave you plenty of latitude to develop your own method of enticing that one big fish to take a shot at your offering. A big advantage of artificials over live bait is that they are always ready and waiting in your tackle box.

It is important to remember that these baits are meant to represent live foods that fish feed on.

Initially artificials were designed to resemble their live counterpart as closely as possible. Although this approach to lure making is still the general rule, many lures today also incorporate an **attractor** feature designed to trigger a response. These features can be as simple as a fluorescent color that you can see far away.

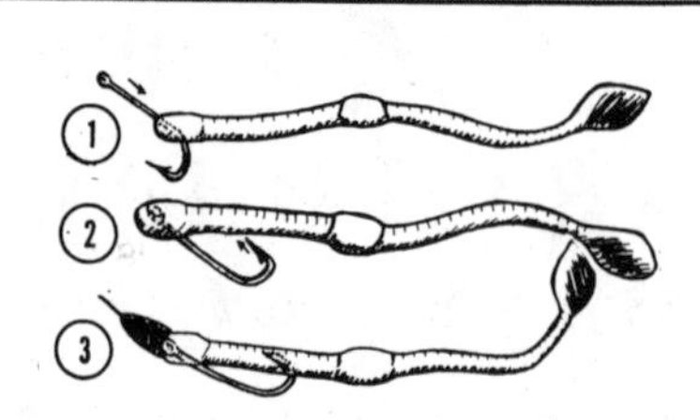

Rigging a worm "Texas Style". 1-Push the hook through the nose of the worm. 2-Pull the eye of the hook out just enough to expose it so that it can be tied to your line. 3-If you need extra weight, add the sinker before you tie your line to the hook. After your line is tied, turn the hook and push the point into the worm.

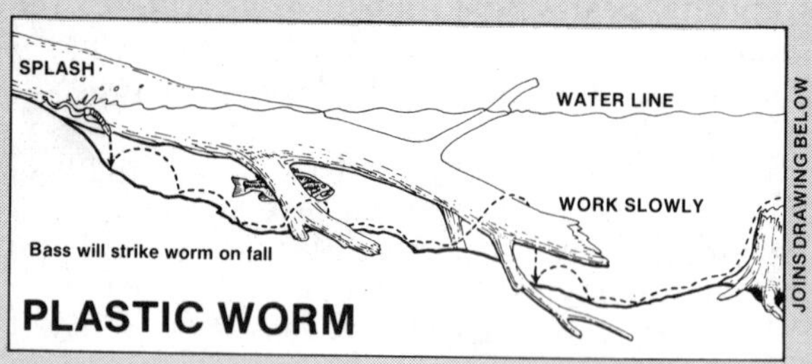

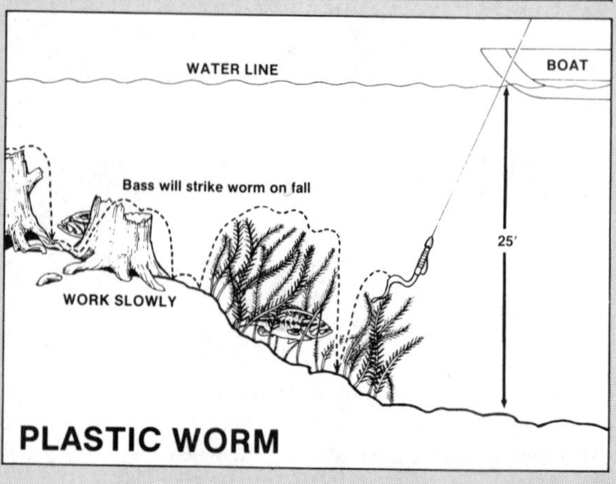

PLASTIC WORMS

The first **plastic worms** were made to look like real worms. They were thin, dark red in color, and had little segment lines molded into their bodies. They were fairly successful. Then the manufacturers decided to try something new and the results have been nothing short of amazing.

Worms began to appear on the market in different colors with soft, fat plastic bodies that squirmed through the water and some even had **twister** tails that gave them a brand new action. A few people laughed until everybody began to catch fish. Now a variety of plastic worms, in many different sizes and colors, is considered a must in any tackle box.

When rigged in the weedless (first called "Texas") style, worms may be fished through the thickest cover without fear of snagging. They may be fished on the surface, using this technique, or on the bottom with the addition of a bullet-shaped slip sinker fitted snug against the head of the worm.

Plastic worms are meant to be fished with a slow retrieve. One of the most productive methods is to cast your rig into submerged brush or downed trees. Using a slow retrieve, pull your worm over each obstacle and then let it fall to the bottom.

There are two schools of thought on when to set the hook when fishing a worm. Some feel you should let the fish run with the worm until he stops and then set the hook. Others say that since the fish sucks the whole worm into his mouth, you should set the hook immediately.

PLUGS

During the season, fish will be found at dif-

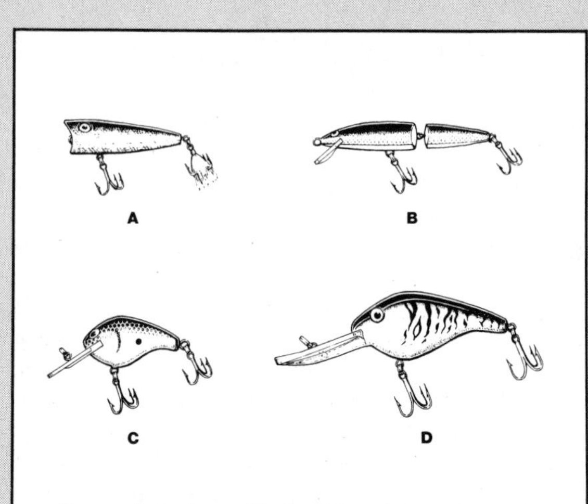

Examples of crankbaits. Plug A is surface plug; plug B a shallow running plug; Plug C a medium running plug; plug D a deep running plug.

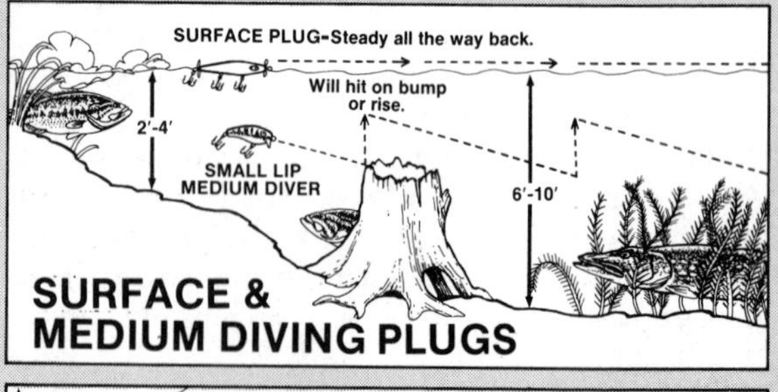

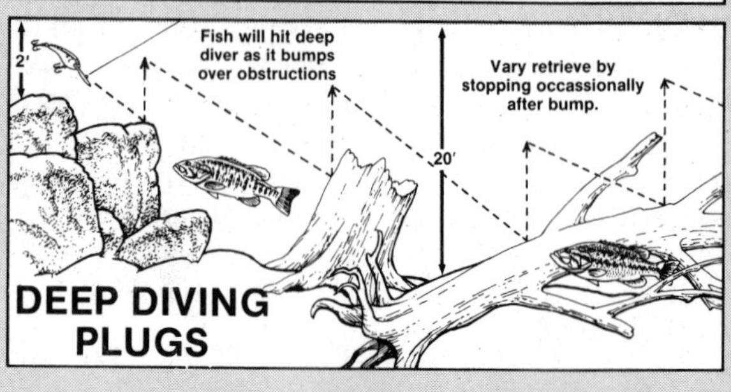

ferent depths in lakes and reservoirs. For this reason, plugs have been designed in four distinct catagories; surface, shallow running, medium running and deep divers. Many of these lures come with a split ring attached to the eye. If the lure you plan to use does not have a ring, attach a plan snap to the eye and your line to the snap, so you do not lose its action.

SURFACE AND SHALLOW RUNNING PLUGS - Surface plugs are meant to imitate injured fish and animals. Cast them out, allow the ripple in the water they create to disappear and try an **occasional twitch** in your retrieve. Shallow runners may be fished as surface plugs or crankbaits that will dive about a foot. Both are normally fished in the early morning and late evening along shorelines and weed beds.

MEDIUM RUNNING PLUGS - When fish are suspended in 8′ to 12′ of water, turn to these lures. Designed with plastic lips molded in or attached, these plugs are meant to imitate bait fish and come in realistic colors and patterns or fluorescent designs you can see a block away. All colors can be effective at certain times, and one general theory holds that light colored lures should be used on bright days, while dark colored lures should work best on overcast days or at night.

DEEP RUNNING PLUGS - Similar in design to medium running plugs, deep divers are sometimes fitted with **built-in** weights to take them to depths of 25′ or more. Some models have diving lips almost as long as the lure itself. These plugs are used almost exclusively in reservoirs when fishing from a boat. When fish are hugging the bottom, this plug can dive to their level.

SPOONS

Though not as popular today as they were many years ago, it is the rare freshwater angler who doesn't have a few spoons tucked away in his tackle box. Spoons are made with and without weedguards. Some even have trailer hooks attached to the main hook. Meant to imitate an injured bait fish, they may be trolled, cast and allowed to flutter down to bottom feeding fish.

One of the most popular styles has a weed guard and a single hook. The hook is often fitted with a strip of pork rind or cut bait used as a trailer. Although the lure can be fished rapidly across the surface of the water, the usual method is a slow to medium retrieve at various depths.

SPINNERS

Spinners are broken down into two distinctly different catagories. The older, single shaft lure is characterized by a blade that revolves around a shaft while it is reeled through the water. The second, newer model, has two arms. One is equipped with spinner blade(s), the other having a molded in jig usually dressed with hair or a rubber skirt. Normally called a **spinnerbait**, this newer model has been further changed in the last few years with the removal of the traditional blade and the addition of a propeller which keeps it on the surface of the water and provides the **buzz** that leads to naming this adaption a **buzz-bait**. This lure is meant to be fished rapidly.

Spinners can be fished at any depth and often the single shafted models have a minnow attached. A good method for determining the working depth of your spinner is to cast it a short distance and notice how fast the lure

Typical spoons with weed guards. Notice also the spinnner blades and other attractor devices that may be added.

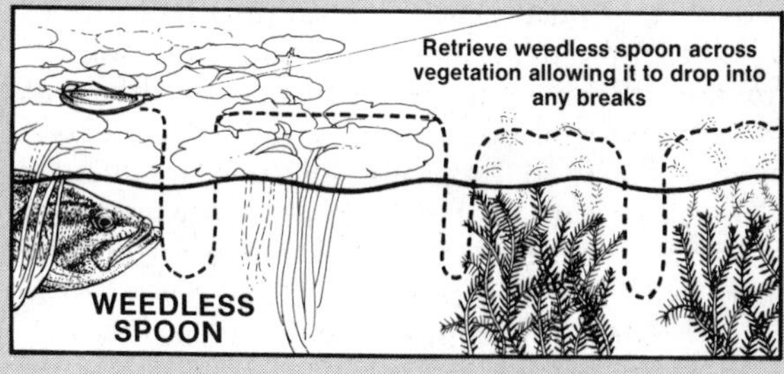

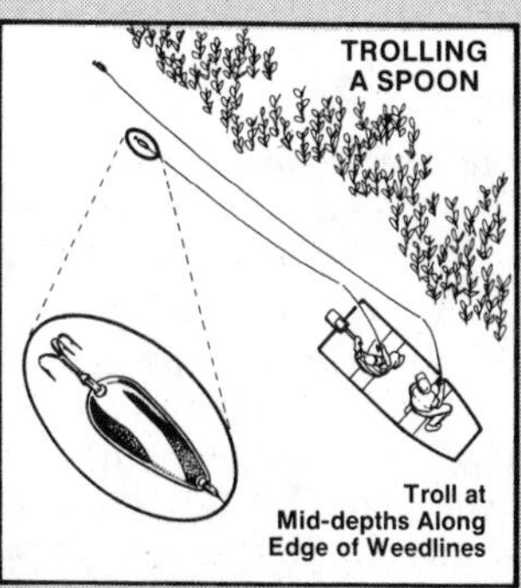

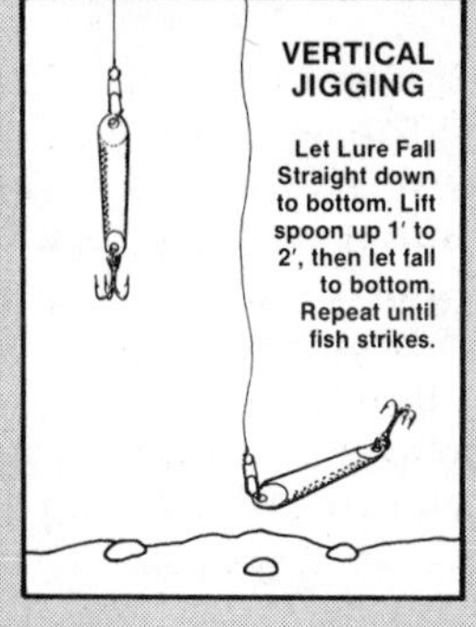

sinks. The average is about a foot a second. With this information you can cast into 10′ of water, wait 8 seconds, and be confident that your lure will be traveling 2′ off the bottom during your retrieve.

JIGS

Unquestionably the most versatile of all artificial lures, the **jig** is simply a hook with a molded lead head, which acts as the sinker. Jigs are normally dressed with bucktail (deer) hair, plastic hair or feathers to form a skirt which extends slightly past the bend of the hook. The name jig refers to the original way of fishing this lure-using an up and down motion of the rod. With this method, the rod tip is swept upward and dropped sharply which allows gravity to **swim** the jig downward.

Jigs are manufactured in an infinite number of sizes and designs. Sometimes the molded heads will vary in shape to impart different swimming motions to the lure. Popular sizes, such as the 1/16, 1/8 and 1/4 ounce, are usually fished with bait such as minnows or worms. As with the plastic worm, jigs can be **weedless** lures with the addition of an arm or with the hook pushed into a plastic twister body.

SCENTS

Before you consider adding one of the artificial scents on the market today to your lure, be sure it is legal in your state. All fish have **olfactory glands** and are able to smell. This sense will vary in importance among the species of fish found in the reservoirs and lakes of this country. Some species, such as the catfish, use scent to locate food while others depend more on sight and sound. While scent will work with

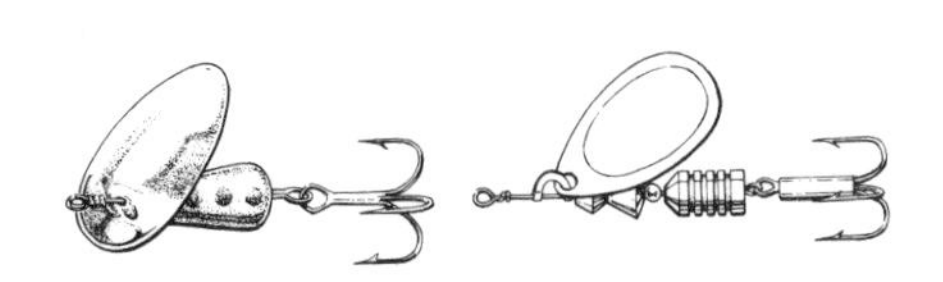

These lures are examples of the single shaft spinners. Many times a minnow is added.

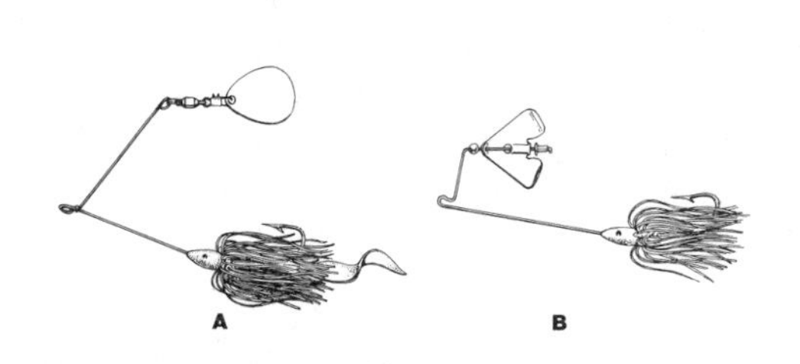

In this illustration, lure A is a spinnerbait and lure B a buzz bait.

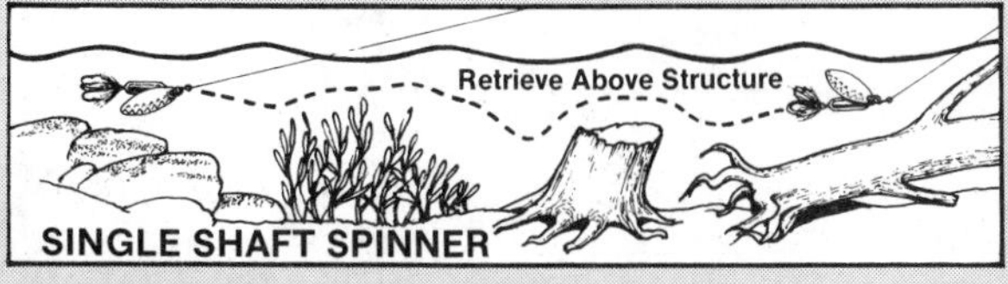

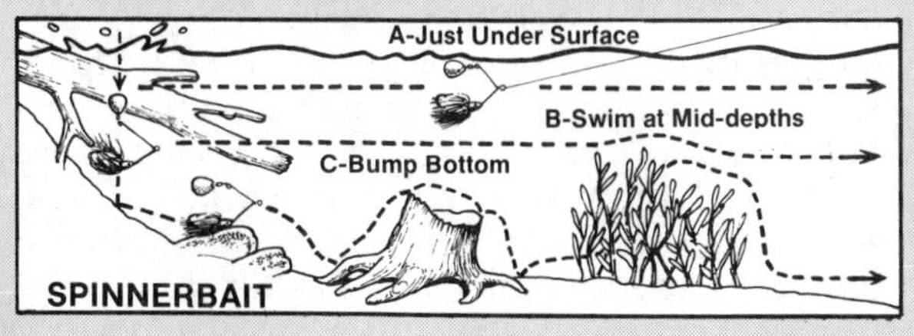

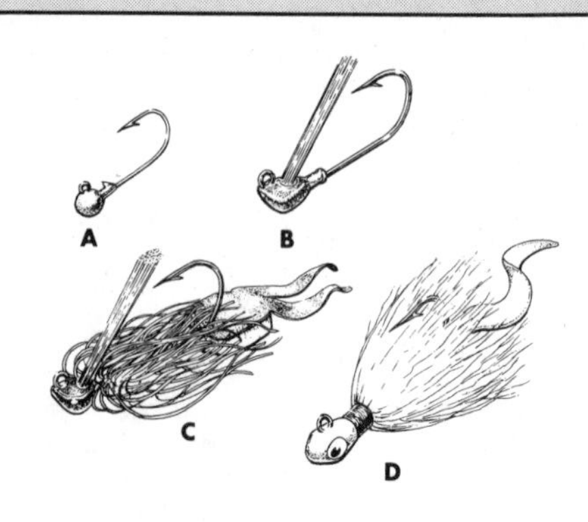

Examples of the versatility of the jig. Jig A is a plain jig; jig B has a weed guard; jig C has a weed guard, rubber skirt and piece of pork rind added and is better known as a pig n' jig; jig D has bucktail (deer hair) and a plastic grub attached.

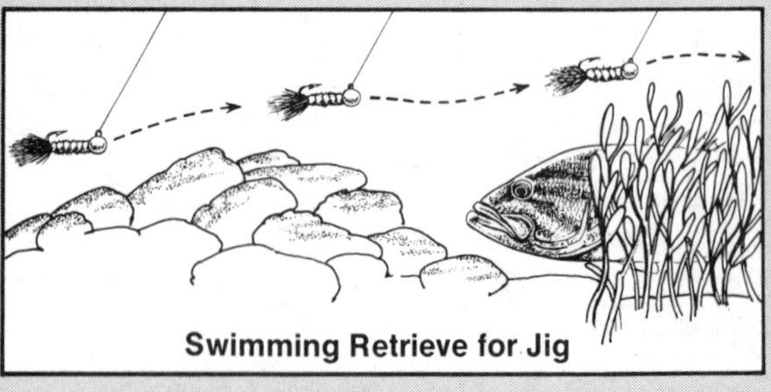

Swimming Retrieve for Jig

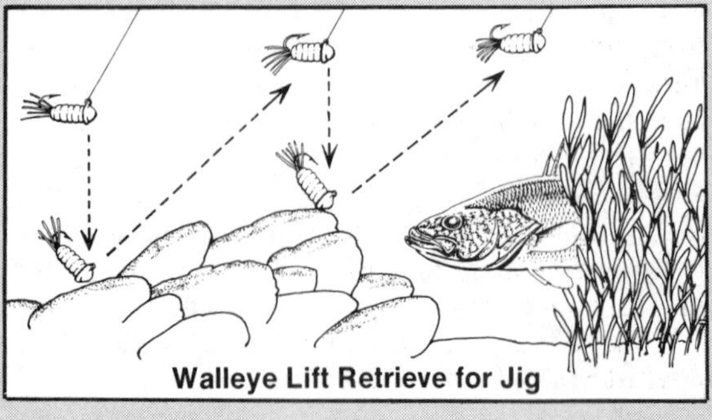

Walleye Lift Retrieve for Jig

any lure, it works best with lures that are moved slowly thorugh the water such as **plastic worms** or any jig with a soft plastic **twister body** attached.

SELECTING THE PROPER LURE

There is no hard and fast rule in lure selection. Certain lures have a better track record in particular situations. For instance, if you come across a fallen tree, the natural choice is a plastic worm. Anything other than a snagless bait is bound to get hung up.

Another clue might be what live bait is working best at the time. Since artificials are meant to imitate live bait, pick a lure that most closely resembles what the fish are presently feeding on.

It makes sense to use a lure that is designed to be used at the depth that the fish are holding. What about color? Those fluorescent colored lures are meant to **trigger** a reflex action. If that doesn't work, try a more realistic color. Ask questions at the tackle shop. On any given lake, for some reason, certain colors work, others don't.

Perhaps the fish you are after are near the surface of the water. Color isn't the only way lures can trigger a reflex. Known as **disturbers**, some lures are designed to create a ruckus as they travel through the water. Sometimes the design change that accomplishes this is a concave nose design, other times it might be the addition of a propeller. The concave nose lures are mainly surface lures, while the propeller isn't necessarily limited to that depth.

Fishing is a continual learning process. That is what makes it so much fun.

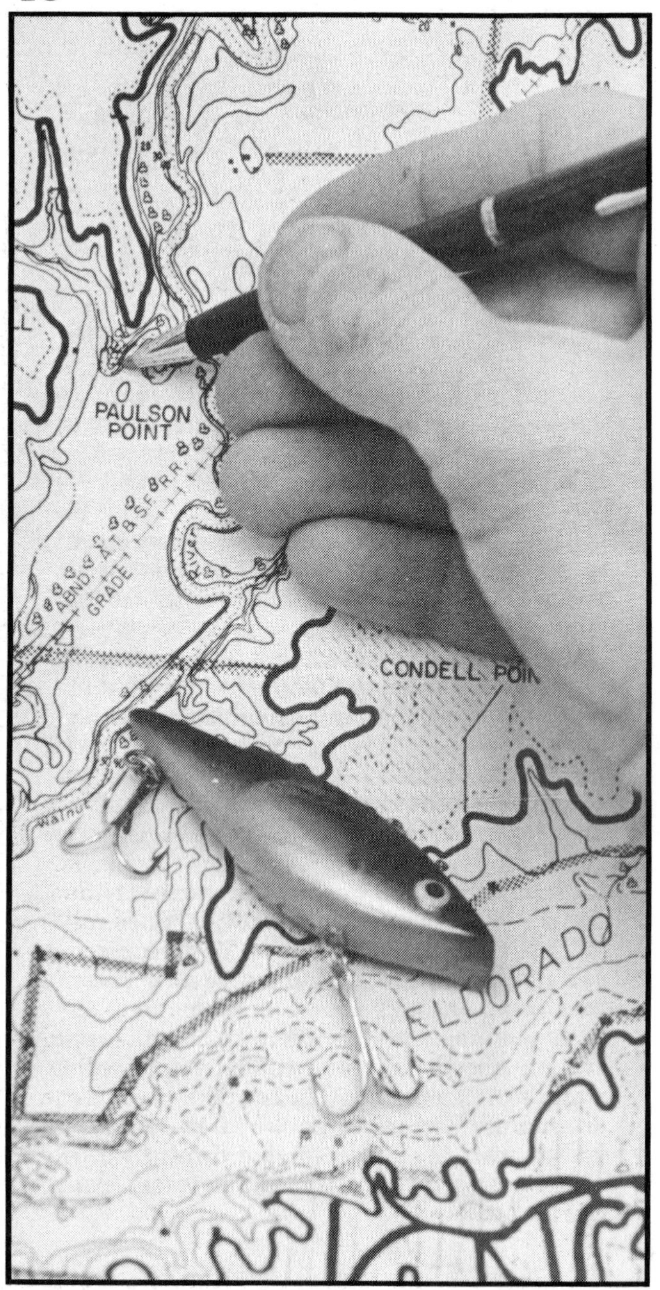
PAULSON
POINT
ABND. A.T.&S.F. R.R.
GRADE
CONDELL POIN
Walnut
ELDORADO

8

READING A RESERVOIR

Lakes and reservoirs are commonly referred to as **still waters**. Unlike rivers and streams, with their abundance of structure above the water line that indicates obvious holding areas for fish, still waters give little indication of depth or bottom configuration.

There are tools you can use, such as structure maps and electronics, to unlock these secrets of the mysterious depths, but sometimes they might not be available to you, or in the case of electronics, not practical when fishing from the shore.

The first thing you should do, no matter if fishing from a boat or walking the shoreline, is observe the land area that leads into the water. If there is a steep, rocky hill that runs to the water's edge, chances are it continues to the bottom of that portion of the lake. The existence of such things as water plants and other aquatic life will normally indicate that the lake is rather shallow in that area, and usually, in most man made lakes, the deepest section of the waterway will be near the dam.

Sometimes there can be too much of a good thing. A case in point could be an over abundance of structure. An area just beyond the shoreline cluttered with many tree stumps will probably hold scattered fish. However, an area with fewer stumps may hold a concentration of fish.

Visual Sightings	Likely Species
A. **Old Road**— Appears to enter lake. Most likely this is an old road bed.	largemouth & smallmouth bass
B. **Break in Trees**— Could be a continuation of old road on other side of lake.	Same as above
C. **Deadfallen Tree**— Often provides good cover for fish.	largemouth bass, crappie, sunfish
D. **Weeds** — may be submerged weeds such as milfoil or floating weeds such as lily pads	walleye, pike, musky, sunfish, largemouth, smallmouth bass
E. **Fruit Trees & Grassy Knoll**— contrasts to forested terrain of rest of lakeside. May be remains of old homestead. Check offshore for foundations.	largemouth bass smallmouth bass
F. **Short Pier**— Typical of deep water.	smallmouth bass
G. **Long Pier**— A shallow water pier.	largemouth bass
H. **Anchored Boat**— Why is boat anchored in middle of lake? Perhaps a submerged island is under it.	walleye, musky, crappie, smallmouth
I. **Island**— Always a good place to check for fish.	musky, smallmouth sunfish
J. **Feeder Stream**— Will draw fish in spring.	all gamefish
K. **Point**— Congregates game fish.	all gamefish
L. **High Bluff Shoreline**— Indicated similar conditions below water. Steep dropoff.	musky, walleye
M. **Gentle Rolling Shore**— Same slow underwater drop.	largemouth, walleye sunfish
N. **Dam Area**— Deep water.	all gamefish

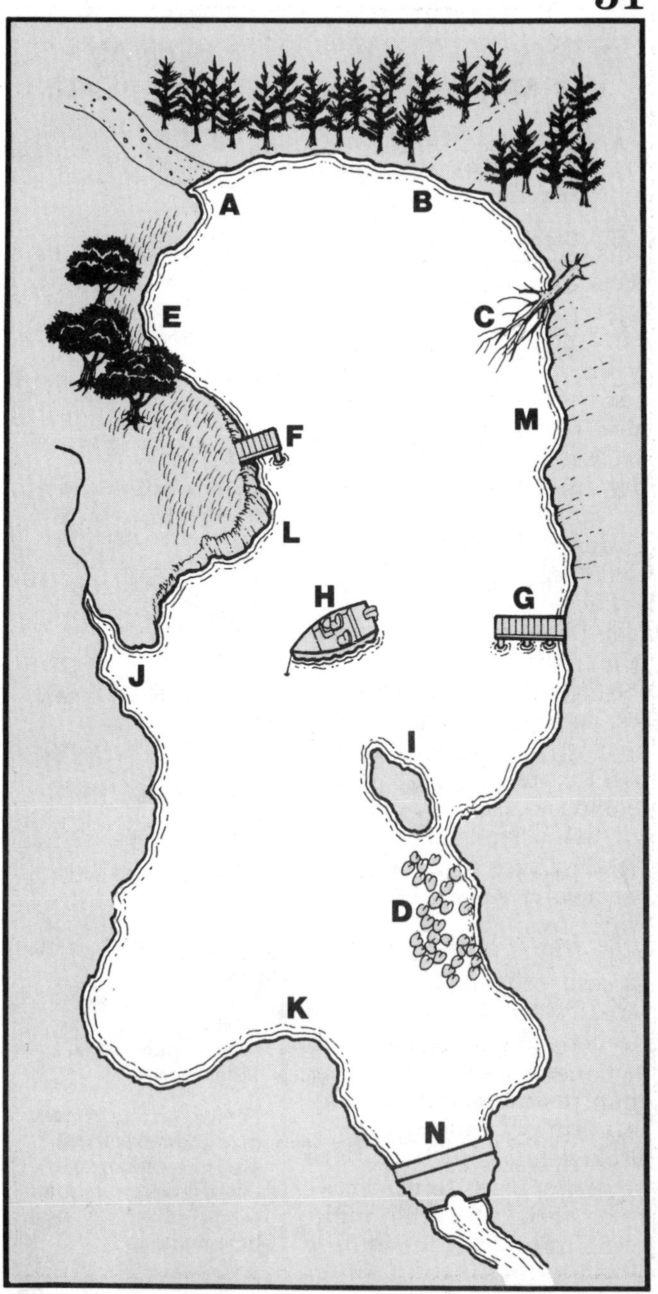
A
B
E
C
F
M
L
H
G
J
I
D
K
N

STRUCTURE MAPS

The **United States Geological Survey** has made contour maps for just about every inch of this country. This is the source for most **structure maps** used by fishermen. If a map maker can find one of these maps before or just after an area has been flooded to make a new reservoir, he has the necessary materials to identify drop-offs, old road beds, old buildings and any other underwater feature that could hold fish.

These maps are important tools for any fisherman, even if his boat is equipped with the latest in marine electronics, because you can't find a spot with electronics if you don't know where to start looking.

Obviously, before you can flood an area to make a new reservoir, there must be water available. This water may take the form of a stream or small river and usually runs through the deeper sections of the eventual reservoir. On a structure map, this underwater feature is normally indicated by a broken line and may have several or many feeder creeks that feed into the main channel. Called **creek**

HOW AN EXPERT READS THIS MAP & WHY

1

First, he checks for the depths at which the lines are drawn. This tells him the depth of the lake and the swiftness of the bottom drop off.

2 & 3

Then he looks for feeder stream(s) and traces path to dam. This indicates original stream bed and deep water.

4-11

Major structure is next checked.
4 is large bar extending from island.
5 is large under water point.
6 is steep drop off.
7 is suspected underwater island.
8 is shallow bar between island and shoreline.
9 is slow tapering drop off.
10 is shallow cove.
11 is deep cove.

Finally, he might mark map (pencil is best) of areas he wishes to fish - then cross check map readings with visual observations.

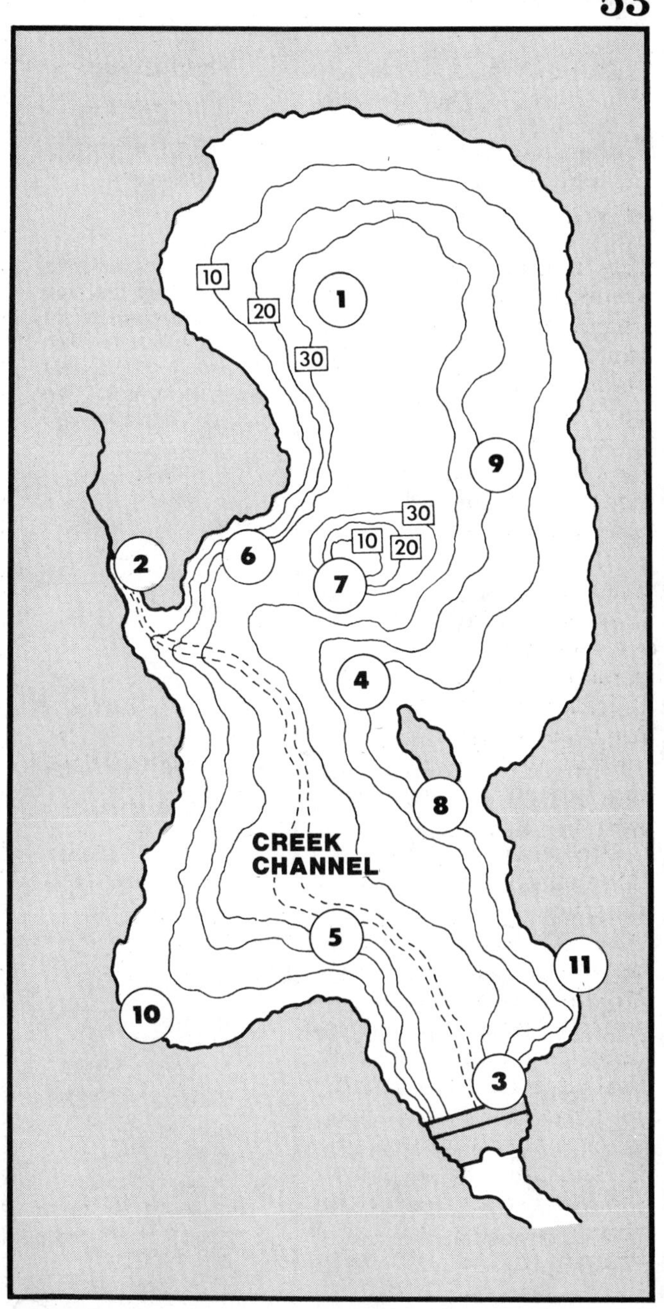
10
20
30
1
9
30
10
20
2
6
7
4
8
CREEK
CHANNEL
5
11
10
3

channels, they are a very important element of structure on any reservoir.

Just as snow will melt first on the **sunny side** of the street, shoreline on any reservoir will warm up first if it has a **Southern exposure**. This is important to remember because these areas will normally contain most spawning areas and produce most of the early season catches.

Coves are always good places to try. They provide protection and good spawning areas.

Feeder streams are also worth a try. Warm water runoffs in early spring could be **just what the doctor ordered** for early season feeding binges.

Long bars, steep points and **spillways** below dams are also areas that should be explored throughout the year. All these locations are indicated on these maps.

As we have learned, different species have very specific likes and dislikes. The trained angler can identify good starting points to search out gamefish with the aid of a map even without ever having fished the lake.

Reading a structure map takes practice and that will be the purpose of the rest of this chapter.

LET'S SEE IF YOU HAVE LEARNED ANYTHING

Here is another map that you can practice on. It is important to understand what you are viewing and this takes practice. The answers are on page 56.

1. Shallow cove
2. Deep cove
3. Island
4. Submerged island
5. Long, tapering points
6. Short, steep point
7. Feeder creeks
8. Dam
9. Deep water areas

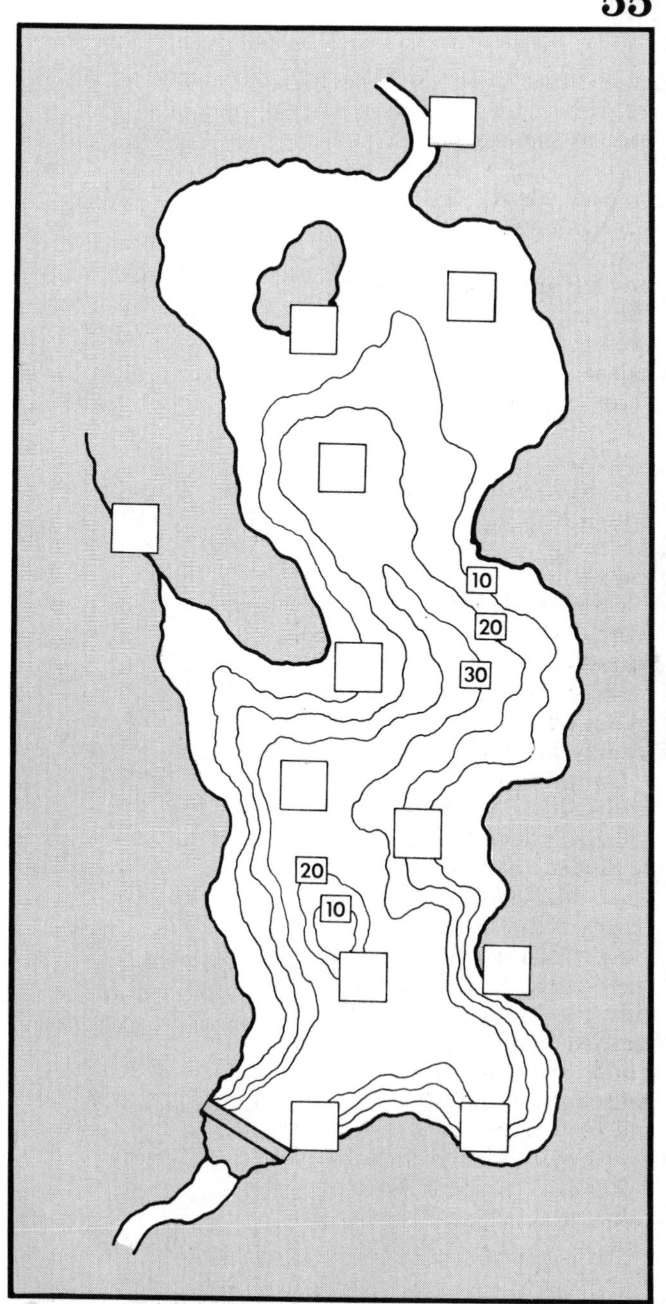
10
20
30
20
10

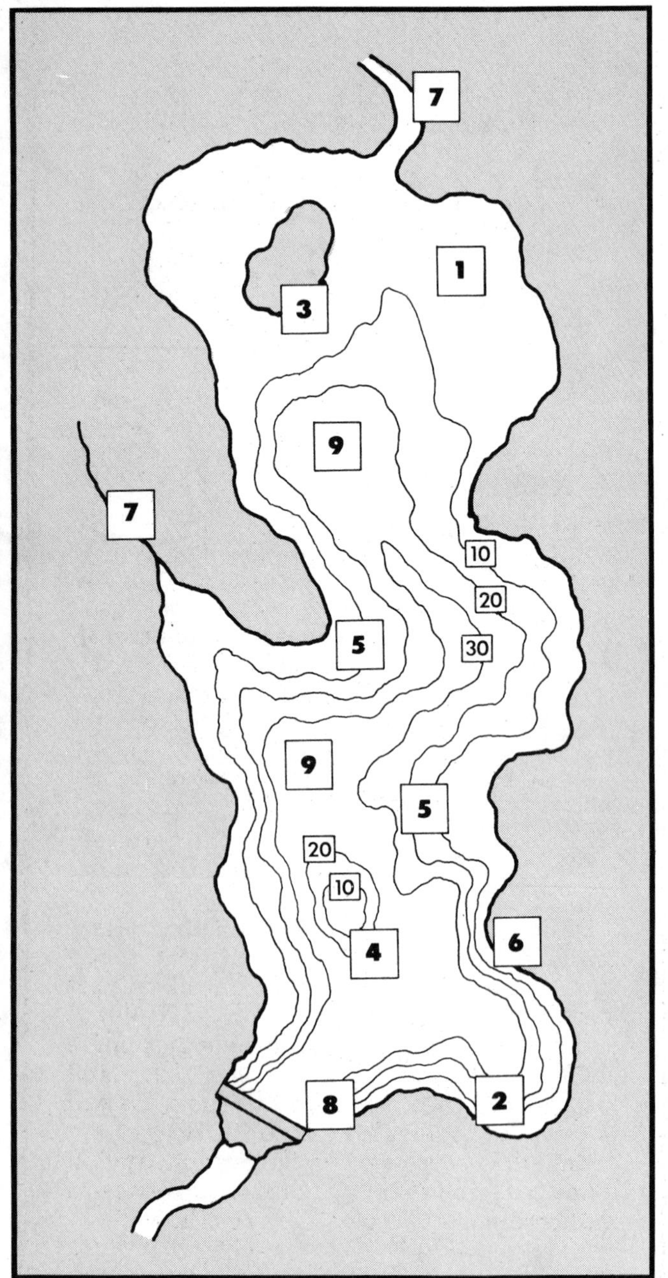
7
1
3
9
7
10
20
5
30
9
5
20
10
4
6
8
2

9

ELECTRONIC FISHING AIDS

Everyone has seen the old World War II naval movies with actors such as John Wayne at the helm of a destroyer in search of the evil German U-boat, with the aid of the ship's trusty sonar. Believe it or not, that sonar unit is the forerunner of today's fishing electronics, or **fishfinders** as they are commonly called.

Employing the same concepts, but miniaturized by the advent of transistors and other new developments in electronics, today's angler has at his disposal equipment that will paint a picture for him of a reservoir's bottom features, indicate if fish are present, determine the oxygen level at different depths, indicate the pH level, measure the temperature and even tell him what color lure can be seen best under varying water conditions.

Despite the advances in technology, the successful fisherman is the individual who understands that these instruments are only part of the picture in consistently catching fish. For, if an angler puts all his learning energies in the electronic gadgetry on the market today and forgets the other basics of the sport, he will never enjoy total success in the activity.

But, this equipment is important, and in many cases represents the difference between a mediocre or truly successful day on the water.

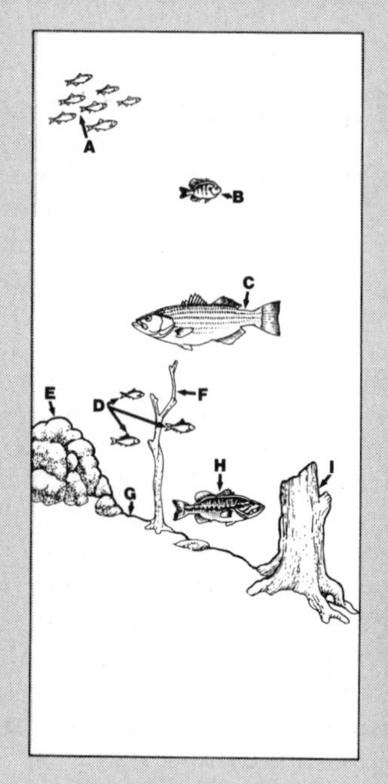

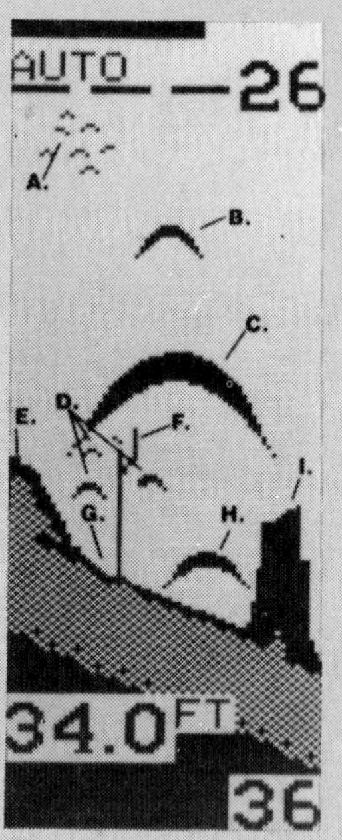

An artist's rendering of what an LCG is portraying on the right. The area shown is from 26′ to 36′. The bottom slope runs from 32′ to 34′. Thus, the total range covered is 26′ to 36′, with the bottom depth at 34′. Each dot represents ½″ in height on the LCD screen.

A. School of Minnows
B. Small Fish
C. Large Fish
D. Minnows
E. Rocks
F. Small Tree
G. Bottom Slope
H. Medium Fish
I. Tree Stump

DEPTHFINDERS AND CHART RECORDERS

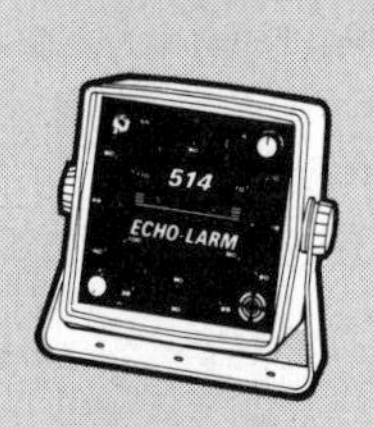

A new version of the old flasher. This one has an alarm system.

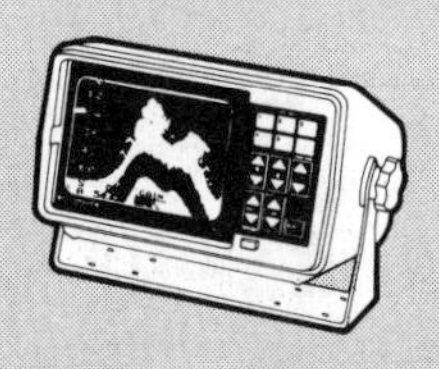

An example of new video systems on the market.

An example of an LCG unit.

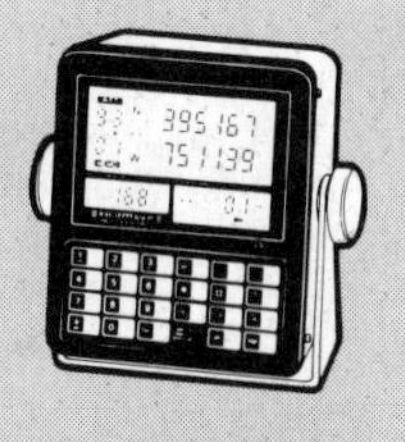

One of new LORAN units.

After the knotted line and weight of Mark Twain's day, the first commercial electronic depthfinders were called **flashers**. These machines had a small light that rotated rapidly around a dial and indicated water depths. With practice, it was possible to determine bottom type and structure that was present in the area.

Technology didn't stop here. Soon **flasher-graphs** were available. Relying on the same technology as the flasher, these units printed out on paper what the flasher saw. They, however, weren't very accurate.

Next came the **straight line chart recorders**. Many feel these units are the best available. Most print out on paper, but video models are available.

Recently, **LCG** units have hit the market. Displaying pictures on their screens using technology similar to a digital watch, one of the main advantages of these units is the ease of seeing the picture in bright sunlight.

With all these units, you get what you pay for.

GETTING THE MOST FROM YOUR SURFACE TEMPERATURE GAUGE

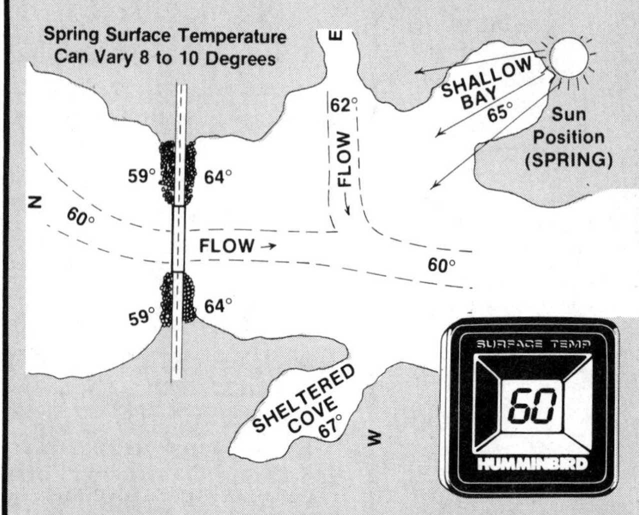

A Surface Temperature Gauge is capable of telling you much more than simply the surface temperature of the water. While this is the obvious function of the unit, when combined with a basic knowledge of fish habits, the temp gauge will increase your skills and boost your day-to-day catch.

Water temperature has a great influence on the movement and activity of nearly every species of game fish. It can govern their migration and feeding habits and probably influences their spawning habits more than any other factor.

In early spring, the surface temperature of a single body of water can vary as much as 8 to 10 degrees (see figure). The position of the sun and its rays, plus the wind direction, can have a definite effect on surface temperature as illustrated. Water in the main channel can be 60° F., while sheltered sloughs and coves might contain 65° to 67° water. Fish such as largemouth bass, crappie or bluegill would like be most active in those warmer areas.

It is also possible for the sunny side of rip rap to be 5 or more degrees warmer than the shaded side. As the illustration shows, the 64° water would probably be more productive.

During mid summer or winter when water temperature reaches the extremes, areas such as underwater springs or feeder streams can vary the surface temperature just enough to cause game fish to increase their activity.

Water temperature can be perhaps the most critical factor in catching walleye. A walleye prefers a range of 39° to 44° water. From ice-out until the surface temperature reaches 55°, your temp gauge can be a valuable tool.

Species such as northerns will react to water temperature differently from other species. Northerns often spawn before the ice goes out or immediately after. They remain active until the water temperature rises above 55°. Then, in the fall, as the water temperature drips below 55°, northerns again become active and your temp gauge once again becomes vitally important.

OXYGEN METERS

As their names implies, this equipment is used to measure the amount of oxygen at various water depths. This is important because oxygen requirements of various fish species varies. Also, if the oxygen count is too low, chances are there are no fish in the area.

TEMPERATURE METERS

There are two different types of temperature meters. Surface units are normally attached to the hull of a boat while the units used to discover the water temperature at varying depths employ a probe on the end of a wire line. Since various species of fish prefer different temperature ranges during the course of a year, this is a good piece of equipment to have.

pH METERS

Similar to oxygen meters, these instruments measure pH on a scale of 0-14. The ideal range is 7.0 which is neutral. Above this number is too acidic, below it, too alkaline.

10

PONDS ARE THE EASIEST

When is a pond a lake, or for that matter, when is a lake a pond. The dictionary tells us a pond is smaller than a lake, yet, the word lake comes from the Greek word **LAKKOS,** which means a pond.

Many anglers ignore ponds for larger lakes and reservoirs. Yet, several state record catches have come out of small farm ponds. These small bodies of water are great early season fishing holes, because, due to their limited depth, they warm up first, causing the fish to become active.

Ponds are very fertile and provide cover and food for the fish. Many of these ponds are owned by your state wildlike agency. Ponds are also found on many farms throughout this country. There was a time when farmers would let you fish their ponds. This is not as true today because anglers abused these privileges and left the area messy. It is important to leave any area you fish cleaner than when your arrived.

The nice thing about most ponds is that they can usually be fished as well from shore as from a boat, due to their size. The following is a general rundown on where to find fish in a pond throughout the year.

Never be afraid to try something new. You can never tell when you might make an important discovery.

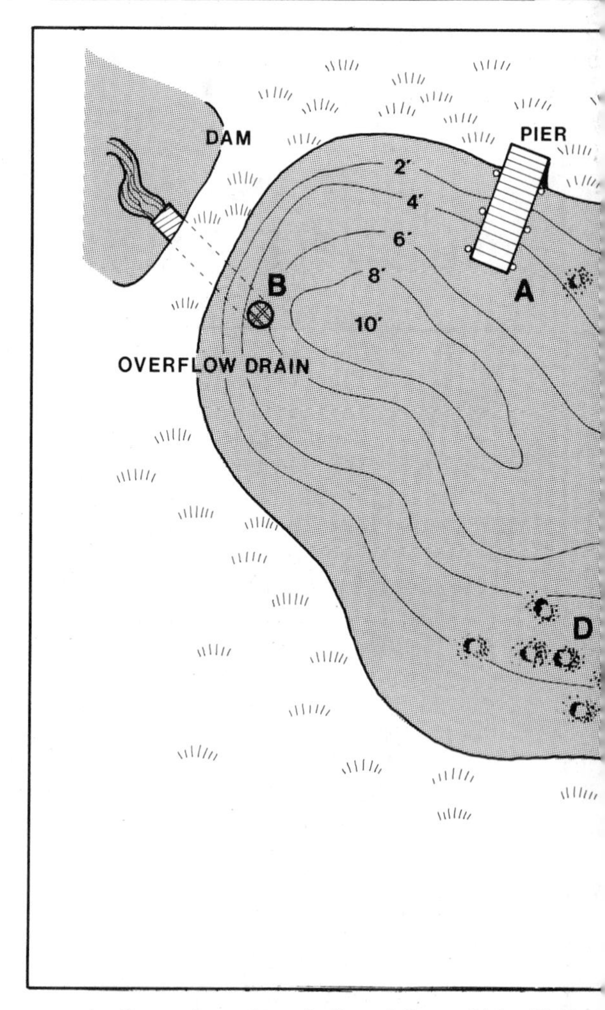

In the early spring, A, B and C would be likely places for bass and panfish. At this time they will leave deep water and begin looking for food and spawning areas. The drain (B) is a good spot because the current brings food to waiting fish. The pier (A) and fallen tree (C) are good hiding places for bass to attack their quarry.

As the water warms, bass will dig out spawning areas (D). Walk along the shoreline and cast parallel

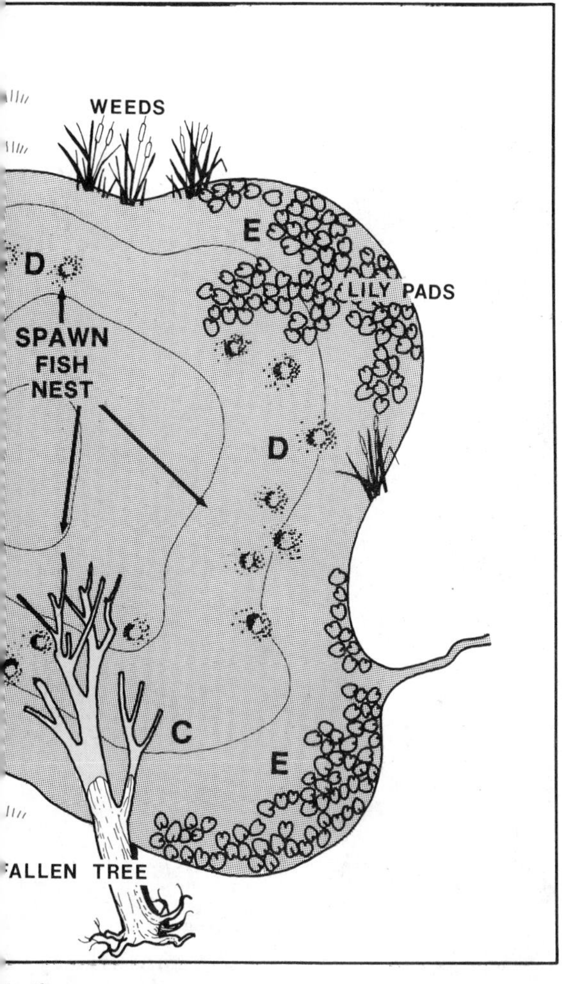

to them.

By summer, lily pads (E) will appear, along with weed beds of aquatic grasses. Fish parallel to these areas.

In the fall of the year, fish will be feeding heavily in preparation for winter. At this time of year, fish return to the areas you found them in the spring. This is one of the best times of year to fish a pond. Try A, B, and C.

LOWRANCE
X-16 COMPUTER SONAR
LOWRANCE

11
UNDER THE WATER LINE

Lakes and reservoirs are probably the most fished bodies of water in this country. The reason for this is simple; just about every community has one. This type of fishing has undergone great change in the last thirty years. However, the basics are still the same. To find fish in these large bodies of water, an angler must understand four basic elements: **temperature, structure, cover** and **depth.**

During the course of a year, most Northern lakes go through two temperature changes that affect the location of fish. Known as **turn overs,** these changes occur when surface water is its densest at 39°F. At this time the water sinks and drives the lighter bottom layer to the surface where it receives oxygen. Fish will follow this turn over and seek proper water temperature and sufficient oxygen supply.

Structure is a general term to describe the physical features found below the surface of the water. It is important because gamefish tend to hold close to these underwater features, mainly for feeding purposes.

Fish favor locations where they can feel safe from predators. These locations could be the same structural features they seek to attack the quarry they are feeding on, or it could be surface cover such as fallen trees.

Depth will usually relate to water temperature. Since warm water cannot hold as much oxygen as colder water, you will usu-

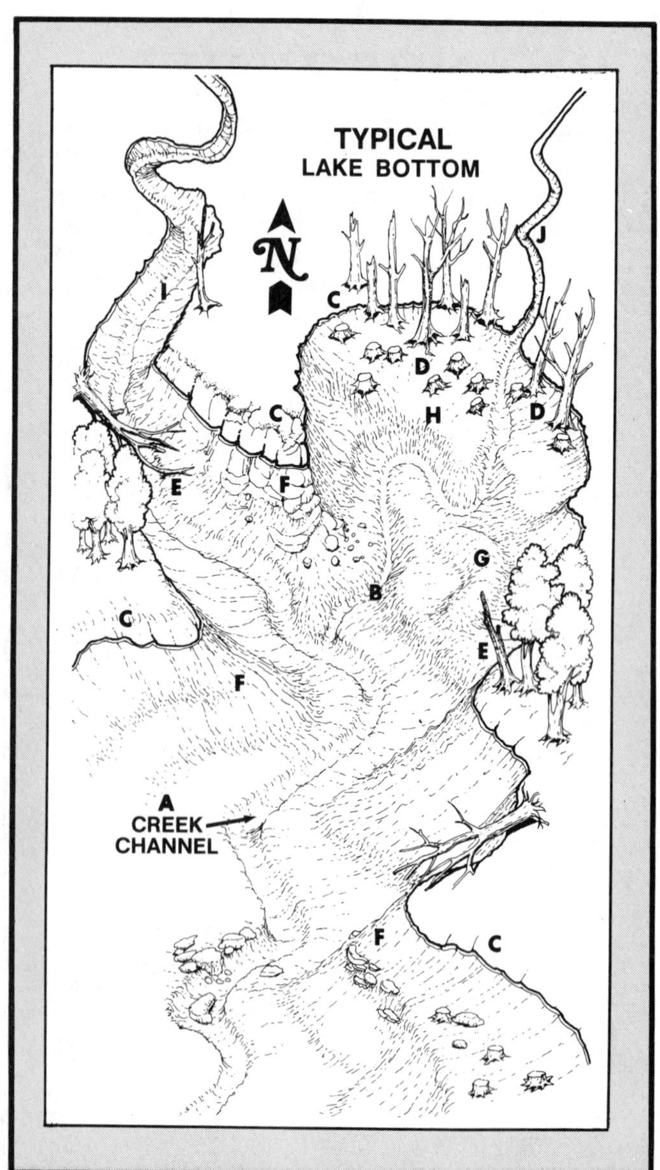

ally find fish at deeper depths during the summer.

STRUCTURE

The illustration on page 68 depicts common structural elements found in most reservoirs. Structure is nothing more than a word used to denote underwater features that various game fish will hold close to, mainly for feeding purposes. A. -CREEK CHANNEL- The original water supply for the reservoir. B. -FEEDER CREEK CHANNEL- A secondary water supply for the reservoir. There may be several of these feeder creeks on a reservoir. C. -SHORELINE WITH SOUTHERN EXPOSURE- Although this feature is above the water line it is important because the northern end (southern exposure) of any lake will warm first and provide most early season action. D. -TREE STUMPS- Good holding area for largemouth. E. -FALLEN TREE- Good largemouth locations. F. -POINTS- Holding areas for most game fish. G. -BAR- Another good area to find fish, especially in early spring and fall. H. -COVES- With feeder stream it will warm early for spawning fish. I & J. -WATER SOURCE FOR RESERVOIR- With warm spring rain runoff, good early season spot.

DROP OFFS

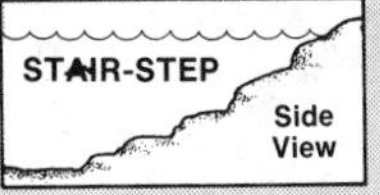

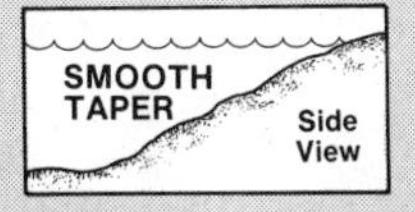

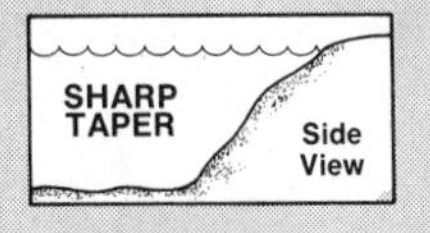

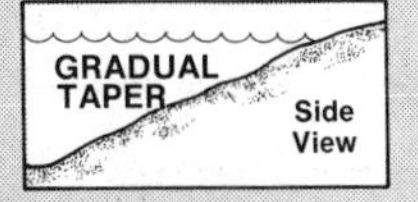

BARS

An underwater ridge. It is higher than the surrounding bottom.

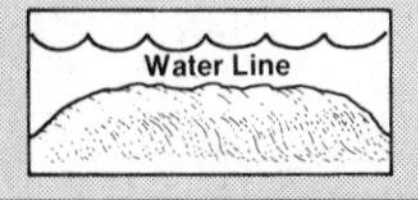

POINTS

A point is when the shore juts out into the water.

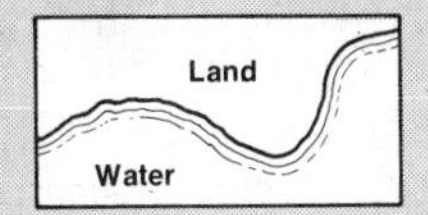

12
UNDERSTANDING YOUR QUARRY

The key to fishing any lake or reservoir is accepting the fact that 80% to 90% of the water **does not** hold fish. Fish are always found in specific parts of a lake during the course of a year and these areas will vary from specie to specie and even among related species, such as the largemouth and smallmouth bass.

While all species of fish must have food, oxygen and the proper temperature range, there will be a variance in need among members of the fish population for other elements such as structure, pH and light levels. Understanding the needs and likes of the specie of fish you are after is the first step in becoming a successful angler.

For some reason, structure oriented fish will establish **patterns** in these waters that can change on a daily basis. The most important element of understanding **pattern** fishing is knowing your quarry. If you have a working knowledge of the places on a lake the species you are after should be found during a certain period of the year, then it is only a matter of eliminating from this list by **trial and error**. In other words, if you finally locate bass holding ing in 15′ to 20′ of water, off a drop-off, and hitting a particular lure fished slowly, chances are you have discovered the pattern and the same will hold true in other parts of the lake with the same underwater features.

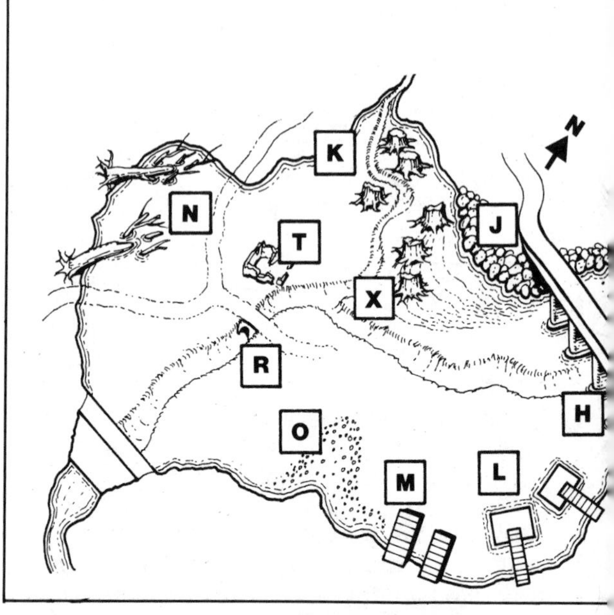

LARGEMOUTH BASS

SPRING 55°-70°, 1′-10′ Areas A; northeast shoreline, and J rip rap getting southern exposure will warm first. Try single shaft spinners and medium running plugs here. Largemouth will seek warm shallows and move to the headwaters B, F & K. Spawning will occur when water reaches 60° F. in same areas. Also, potential spawning spots are P-a backwater, C-lily pads; X-a stumpy point; M-piers, and N-dead fallen logs.

Good springtime lures are spinnerbaits, minnow shaped plugs, and lizard shaped soft plastics.

Following the spawn, spent largemouth bass will go to deeper cover in the same areas. Immature males may continue to be caught near shallow structure.

SUMMER 70°+, 10′-25′ As water gets warmer and sun light penetrates more, largemouth will go to deeper structure. Unlike smallmouth, they prefer area with cover, like stumpy areas X, E, I or standing timber-D the deep weed line-W and drop offs at old road bed-R and foundation-T. They will, however, make brief feeding forays into shallow cover, particularly at dawn and dusk.

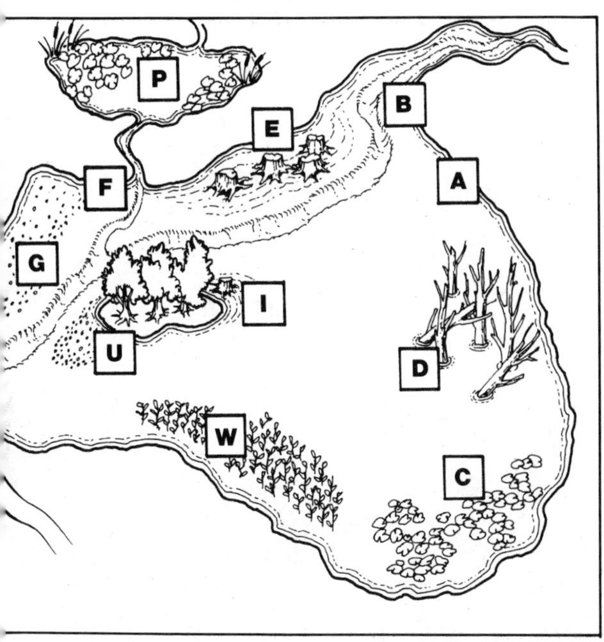

FALL 70°–, 5′-25′ Following turnover, water temperature is equal everywhere. Largemouth may be anywhere and everywhere and frequently gather in groups. Fish cover in the 5 to 25 foot depths.

SMALLMOUTH BASS

SPRING 55°-65°, 7′-15′ Smallmouth, like largemouth, will seek warming water in spring. They will seldom venture into shallow water less than 7 feet deep. Smallmouth bass spawn on clean gravel or sand such as island bar-U, sand or gravel flat-G, and point O.

Early season lures include medium and deep running plugs, jigs and single shaft spinners.

SUMMER 65°+, , 15′-30′ Clean gravel and sand bar U point 0 and flat G are home to smallmouth bass all year long. In summer, they only move deeper.

Other good deep water haunts are bridge pilings-H. and floating piers-L. Jigs catch summer smallmouth because they are small in size and sink deep. Try big lipped, deep running plugs, spoons, and spinners.

FALL 65°–, 10′-30′ Like largemouth, smallmouth may be anywhere in their range. They may be shallow, or deep, maybe both.

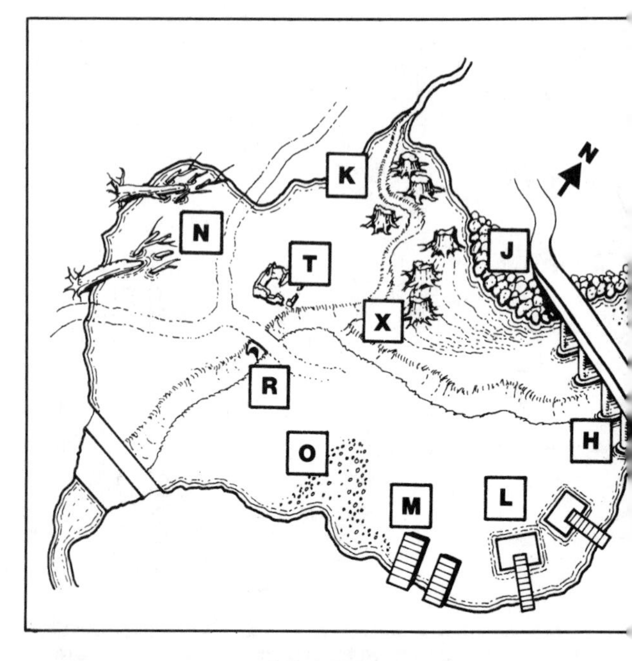

CRAPPIE

SPRING 55°-70°, 1′-10′ Crappie will migrate towards feeder creeks and headwaters-B and K as warming spring rain brings lake temperatures up. When water reaches the lower 60's, crappie will choose shallow, brushy (preferably) spawning sites along lake border. Stumpy flats-E, stumpy bay-K, shallow deadfallen trees-N, shallow piers-M, shallow standing timber-D, and shallow back waters-P, are likely spawning areas. Crappie may even spawn in lily pads-C and forks of shallow branches of deep standing timber-D in some lakes!

SUMMER 70°+, 10′-30′ Hot weather crappie are found in deep water. They will suspend in 10 to 30 foot depths anywhere in the lake, but in relation to old stream bed. Best places to find summer crappie are: deep bridge pilings-H, deep standing timber-D, deeper deadfallen trees-N, and deeper stumpy areas-E, H, and I.

FALL 70°-, 10′-30′ Fall crappie are different than summer crappie in one respect: their appetites are heartier. In autumn, a lake's crappie population puts on a

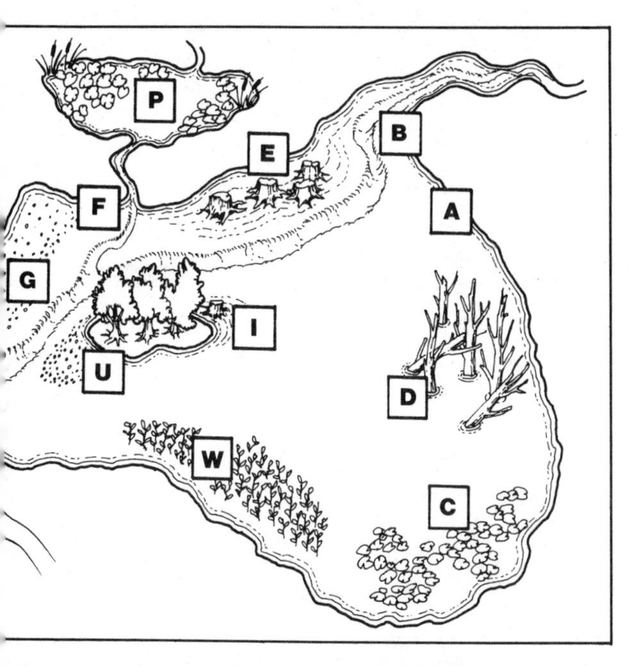

feeding spree which lasts until cold water. They're found in same areas as summer.

SUNFISH

SPRING 55° -70°, 1'-7' Check for shallow sunfish activity in headwater area-B and feeder creek-K. For spawning, sunfish prefer shallow, sandy bays and shoreline pockets where they will build nests, often in great numbers. In addition to areas of water flow already mentioned, potential egg laying areas are: C lily pad cove, and shoreline pockets above G and N. Sunfish may also spawn near pier pilings-M and backwater area-P.

SUMMER 70°+, 10'-20' Much like their cousins, largemouth bass, sunfish will abandon the shallows of spring for the depths of summer. Stumpy, brushy cover in 10 to 20 foot depths, such as: stump flats-E, stumpy points-I and X, deeper deadfalls-N. Deeper bridge area-H and deep weeds-W are also likely hot weather habitations.

FALL 70°–, 5'-20' In fall, sunfish may feed at spring depths, but are more likely to be deeper. Try both.

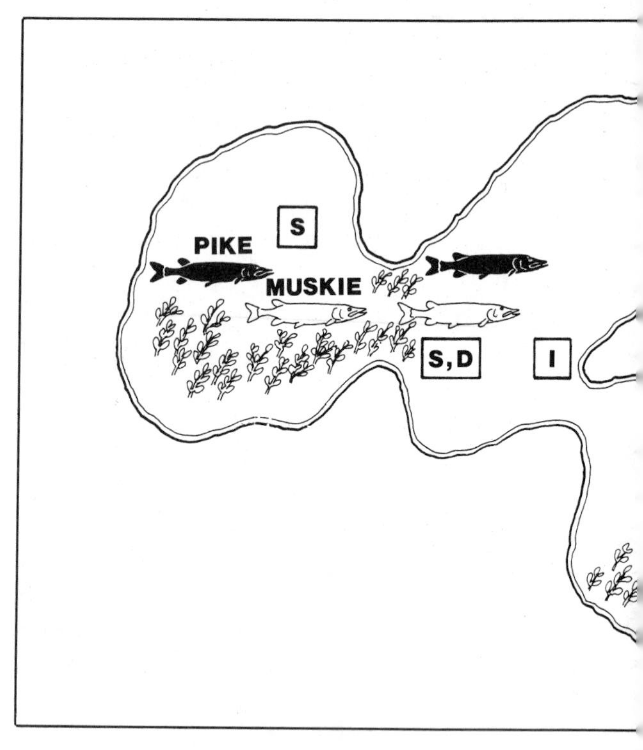

NORTHERN PIKE

SPRING 39°-65°, 1′-6′ The first of the game fish to spawn, northern pike will spawn soon after ice out in shallow, protected bays-S.

Northerns will spend the rest of spring in and around emerging vegetation (weeds) feeding upon available forage species such as yellow perch, crappie, cisco, and suckers.

SUMMER 65°+, 6′-20′ During summer, northern pike will continue to occupy weed beds S and D. However, they will be deeper - along the deep weed line SD or between the weed line and drop off.

FALL 65°–, 6′-20′ As water cools in autumn, and prey species head to wintering areas in deep water northern pike will follow them there. Begin by trying for them in the shallower, weedy areas, S and D, progressively fishing deeper as the game fish migrate.

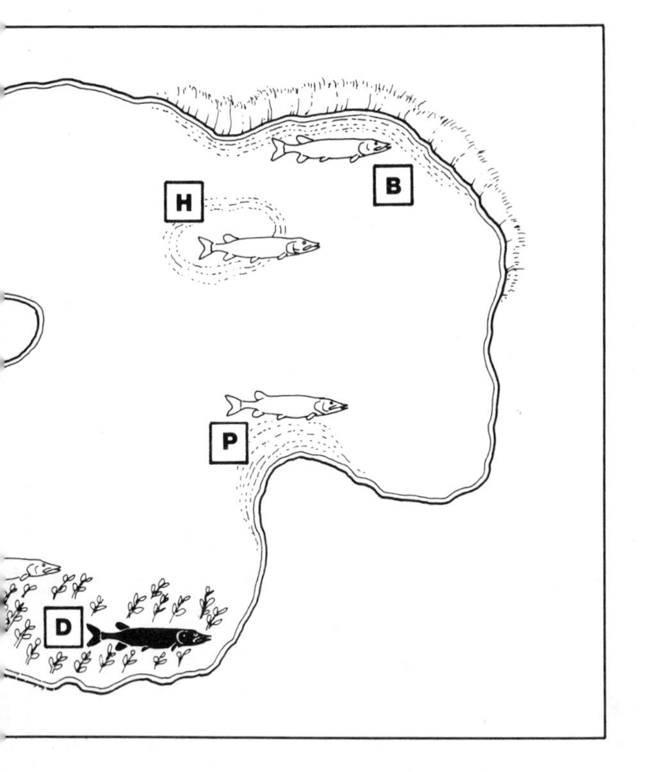

MUSKELLUNGE

SPRING 45°-65°, 2′-10′ Muskellunge spawn after northern pike, walleye, and yellow perch at night in shallow protected bays-S. Like pike, they will spend the rest of spring and early summer in and around weed beds feeding upon forage species.

SUMMER 65°+, 2′-25′ During summer, muskies will migrate to the deeper weeds and weed lines, such as SD. Summer muskies also favor drop-off areas P and parts of a lake which funnel down, such as the area between island and the shoreline point-I, Humps-H in 15 to 25 feet of water.

FALL 65°−, 4′-25′ Fall muskellunge will, like pike, follow shallow water forage fish into their winter holding areas. Steep, bluff banks such as B are also good late season musky spots.

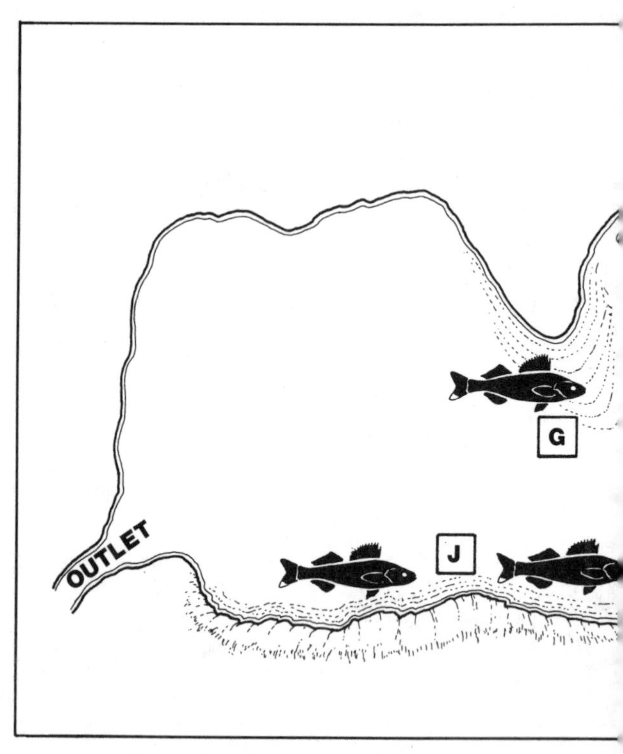

WALLEYE

SPRING 40°-60°, 2′ (night), 15′ (day) Before water temperatures reach 45°, look for walleye near inlet streams-A, flats-B, and submerged islands-C. During the day, the fish should be on the drop off side of structure. At night, they will move up on top of structure.

When water reaches 45°, walleye will be ready to spawn in areas abundant with small rocks - ideally in a wind and wave swept area like area D.

Following the spawn, walleye will move deeper. Submerged islands close to spawning grounds-C, areas of emerging weeds-F, and long tapering bars-G are good places to seek them out.

SUMMER 60°+, 10′-25′ The key to finding summer walleye during daylight hours is determining where the shade is. In shallow, weedy lakes, the only available shade may be in the weed beds-F. In deeper lakes,

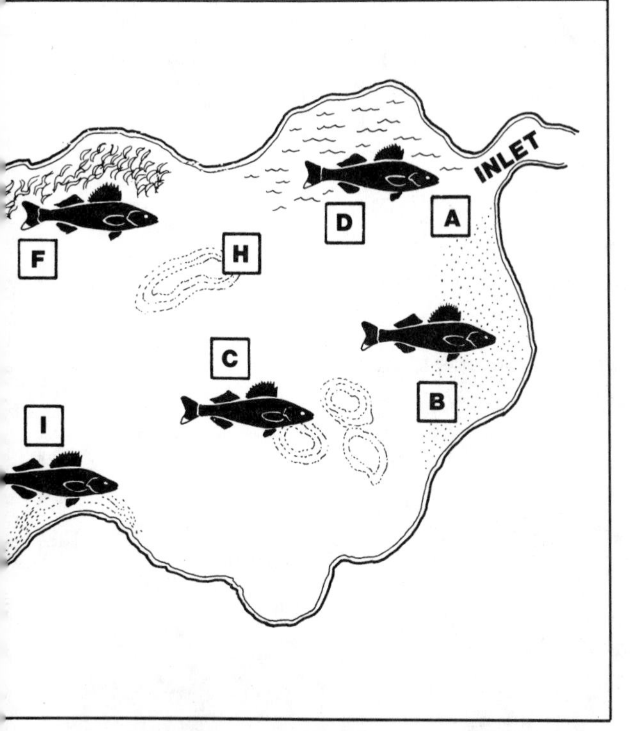

walleye will find shade on the shady side of points-G and submerged islands-C. In larger, deep lakes, summer walleye may leave a smaller submerged island close to spawning areas in favor of large ones further out in the lake-H. There is a limit to how deep walleye will be found, however. That limit is the thermocline. Don't fish any deeper.

At night, summer walleyes may still be found in shallower water.

FALL 60° -, 10'-40' With the fall turnover comes homogenization of lake waters and the destruction of the thermocline. Now walleye roam in search of food. They may be found at the deepest parts of long tapering points, and at the deepest submerged islands during clear, sunny days. On cloudy days and at night, look for walleye on shallower structure. Later in fall, walleye will favor sharper, steeper dropping structure such as point-I, and steep shoreline-J.

Johnson
25

13

SAFETY

Chances are you considered skipping this chapter. After all, what are the odds that you would ever find yourself in a life threatening situation. The answer to that question is you **never can tell**. If you ever do, this will be the most important chapter in this book.

Fishing on lakes and reservoirs is not a dangerous activity if you are properly equipped, take certain precautions and use a little common sense. Certain weather conditions and parts of most lakes should be avoided **at all cost** because of the risks involved. Knowing about these dangers and how to avoid them are necessary for safe fishing and boating.

Safety around water starts with the clothing you wear. While ordinary clothing will suffice, it should be sturdy and serviceable. Because of the increasing dangers of skin cancer, it's a good idea to cover as much of your body as possible. Unless the fish you are after require camouflaged or dark clothing to conceal your presence, it normally is a good idea to wear light colored clothing to help reflect the sun. Always have a jacket or parka available. Remember, you can always roll your pants legs and sleeves up, if desired, for short periods, but this option is removed if you are wearing short pants and a short sleeve shirt.

Wearing a hat or cap is a good idea. In cold weather the addition of a hat will prevent the loss of body heat while in warmer weather it will help guard against sunburn and assist in eliminating glare off the water.

Gloves are a good idea in cold weather. Although it might seem incompatible to wear gloves while fishing, the fact is that gloves are available which are pliable enough to allow you to tie knots and cast accurately.

Sun glasses are necessary outdoor gear. Not only do they protect against glare but shield your eyes against ultraviolet rays that many experts feel can damage them.

Insect repellents and sun blocks are also useful items to have with you. The best repellent is made from a substance called DEET. But, be aware that DEET will damage any plastic item. Great care must be taken when using this lotion since plastic is used in the construction of your tackle box, lures and parts of your reel. Sun block is necessary to prevent burning. Number 15 or higher is recommended.

Weather conditions should also be taken into consideration. **Lightening** can be very dangerous. This is especially true if you happen to be out in the middle of a large lake during a lightening storm. If you get caught in one of these storms, get off the water as soon as possible. If you are trapped on the water, lay down all fishing rods and take down any radio antennas. Graphite rods are especially dangerous as graphite will attract lightening. Also, sit or lay on the deck of your boat so you are not an elevated target.

Even a peaceful lake can be dangerous in storms with **high wind** conditions. If you ever find yourself in severe wave conditions, face your boat, and ride into the waves until you can get off the water as soon as possible.

The section of a reservoir close to the dam is most always **posted**. Normally, not only is it against the regulations to go into that area, but foolish to even think about doing so. This is the deepest part of the reservoir and the area affected if the dam is opened. When this happens, the current created by the release of water over the dam can suck your boat into very dangerous conditions.

PFD'S, or personal flotation devices, are required when boating by the Coast Guard or, in the case of lake and reservoir boating, by your state. The regulations generally require that there be one wearable PFD for **each** passanger and a throwable ring or cushion PFD for boats of a certain size or features. This safety gear is rated by type. Although lakes and reservoirs are not considered as rough waters requiring the usage of Type I, this gear provides maximum flotation and will turn an unconscious person head and face up to prevent drowning. It is reversible for quick wearing and is available in adult and child sizes.

Type II PFD's will also turn the wearer to a vertical and slightly backward position to prevent the unconscious from drowning. They come in three sizes but do not give the long term protection and flotation that Type I does.

Type III PFD's are the type usually used by fishermen. They frequently have pockets and can double as fishing vests. They will maintain a face-up position, but will not turn an unconscious person to this position. Many sizes are available in both vest and jacket styles.

Type IV PFD's are designed to be grasped, not worn. They include flotation cushions and rings, and are designed to be thrown to a person who is overboard.

PFD's also help to maintain body heat in the water and help in preventing or minimizing hypothermia.

To prevent accidents and make boating enjoyable and safe, several rules of the road must be observed. An operator is considered in violation of boating laws if these are not followed. Negligent operation includes speeding, operating a boat in protected or swimming water or slow speed areas, skiing too close to swimmers or fishermen, causing waves to other boats or docking areas, taking the right away from a sailboat, running at night without lights and operating under the influence of alcohol. It is important that you are not only aware of these laws, but make sure that you practice them. They were designed to save lives and make boating safe and enjoyable.

All fishermen should know how to swim. As you probably won't have a lifeguard available on the lake you are fishing and if the regulations say it is legal to do so, try to follow the following precautions, if you decide to take a dip.

If you are by yourself **don't** go for a swim unless you fall overboard by accident. If you have a partner, make sure he or she is an **experienced** swimmer. It goes without saying if you have any doubts about your ability to swim, stay out of the water. Stay out of cold water, since hypothermia can result.

The word **hypothermia** has already been mentioned twice in this chapter. It's a good idea to understand exactly what hypothermia is and what to do if you encounter someone suffering from it.

Hypothermia is the chilling of a person **beyond** that person's **ability to rewarm** the body. If not corrected, chilling of the body core temperature causes weakness, hallucinations, uncontrollable limbs, finally unconsciousness and death. Part of the body's protection is to sacrifice the blood flow and maintenance of body heat in the limbs to protect and maintain body heat for the internal organs and brain.

One fallacy is that hypothermia can only occur in cold water. In fact, it can occur in almost **any temperature**, once the body begins to get cold or chilled through rain, wetting or wind.

To prevent hypothermia, dress warmly, carry spare clothing, keep your head warm, and use a waterproof parka or rain gear to protect against loss of body heat by rain and wind.

If you encounter someone who has succumbed to hypothermia, try to **secure professional medical help** as soon as possible. While waiting for this help or if no help is available, completely remove their wet or cold clothing and replace with warm clothing or place them in a warm bath. The administration of hot sweet drinks is approved but **no alcoholic beverage** should be used. **Do not** warm the limbs and **do not exercise** the patient by walking. **Warming the limbs** will often cause increased blood flow to the limbs, resulting in **stroke, heart attack, and death**. In **extreme** cases, a warm bath, with the arms and legs outside of the bath water is ideal. A **good substitute** is a warm sleeping bag or electric blanket (only on the trunk, abdomen and head). Hypothermia is not a casual concern and each year many fishermen and other outdoorsmen die from it.

A person **overboard** can be a very serious situation. Many times it results in as much panic onboard as it does with the individual who fell overboard. It is **important** for all concerned to remain calm. The **throw-row-go** method of life saving is the recommended procedure to use to assist the individual in trouble.

The first safety procedure is to **throw** a life saving device to them. This should be a type IV PFD such as a ring or cushion, preferable attached to a rope. Lacking this, any floating object such as a cooler will help. If this is not possible, **row** a boat to the person, preferably with an assistant in the boat to help pull the person over the stern (rear) of the boat. If using a motor powered boat, the motor **must** be turned off before reaching the person overboard. If not, you run the risk of cutting the person with the motor prop. Do not allow the person overboard to try to pull themselves into a small boat over the side, since it might overturn your boat. Only an individual experienced in life saving should attempt to **go** into the water after the person. Drowning people will often panic and can hamper, injure or even drown their rescuer.

It goes without saying that **liquor and boating** don't mix. The heat of the sun, glare from the water and rocking of the boat only **increase** the intoxication process. Safe boating requires a clear mind and common sense. Being intoxicated **eliminates** proper usage of these abilities.

Your car travels on tires over roads where any potential hazards may be clearly seen. A boat travels on its hull, that is submerged, in the water where potential hazards, in many cases, can not be seen until it is too late. These potential dangers can be greatly reduced if you are traveling at a **safe**, **reasonable** speed.

Knowledge of basic first aid is also essential on any fishing trip. Normally, you are in an area where medical care is unavailable and many times action should be taken before this help can be secured.

The most common medical problem that anglers face is the removal of a fish hook that has penetrated the skin beyond the barb. Usually it's a good idea to leave the removal to a professional if you can get to a medical facility in an hour or two. **In no cases** should you ever attempt to remove a hook from around the eyes, from the face, from the back of the hands or from any area where ligaments, tendons or blood vessels are visible.

The simplist removal is to cut free the rest of the fishing lure and use a loop of heavy twine (heavy fishing line is satisfactory) around the bend of the hook. Next hold down the eye and shank of the hook, pressing it lightly into the skin. Grasp the loop and with a sharp jerk, pull the hook free. The downward pressure on the eye and shank of the hook clears the barb and allows it to travel out through the puncture wound. Be sure to have a **tetanus** shot as soon as possible unless such protection is already in effect.

Small cuts can be handled with antiseptic and bandages. Larger or deeper cuts require pressure directly on the wound to prevent excessive bleeding. To do this, use sterile, sealed gauze pads or as an alternative, an unfolded clean handkerchief. In the case of severe bleeding, in which an artery or vein has been cut, a tourniquet may be necessary as a last resort.

For cuts on arms and legs, the best direct pressure or tourniquet position is at the joint immediately above (closest to the body) the cut where the major blood vessels travel over or near the bone. Use direct pressure here, or a properly applied tourniquet, to stem the flow of blood. In either case do not apply too much pressure, or pressure for too long. As soon as it is possible, call a doctor, get the individual to a hospital or call paramedics.

Fishing was never meant to be dangerous, just memorable. A comfortable, safe trip is far preferable to an uncomfortable, dangerous trip. Many times the difference revolves around a little common sense.

14

ETHICS & OUR ENVIRONMENT

Fishing is an extremely **personal** pursuit. An angler may opt to fish passively, sitting on the bank of a creek, pond, or lake. Another fisherman may take a more aggressive approach and actively stalk his quarry. The level of participation chosen is not something to be **judged by others**. It is an individual decision in the search for enjoyment, relaxation, and delicious, healthful food.

Each enthusiast, however, must shoulder a **responsibility** for the resource. There is a growing necessity to determine one's own fishing ethics aimed toward insuring that the same opportunities will exist for the future generations. This is a personal commitment that must be followed diligently without policing.

For some, it goes beyond the attitude afield or the respect shown to others. This commitment is **much more** than simply obeying established rules and regulations. It may foster individual efforts to leave the aquatic environment in better condition than it was found or it may encourage an angler to become part of a group battling for the future of the resource.

No matter how ethics and commitment manifest themsleves, they must be addressed by everyone who plans to participate in fishing and use the aquatic resources of this great land.

Unfortunately, the preceding ethical standards are not being met by many fishermen in this country. As our population expands, the amount of fishable waters in this country is going in the opposite direction. The reason for this unfortunate situation is simple. A great deal of the surrounding shoreline of the impoundments in this country are privately owned. For years anglers could easily obtain permission to fish these waters from the landowner. But, little care was taken by these people to **clean up** after themselves. It didn't take long for the landowners to become fed up with the situation and **post** their properties. These problems also are of great concern to many state officials charged with protecting state lands. Don't be surprised if selected state lands are posted in the future.

It is important that you **always** have a trash bag or other receptacle with you. It is more important that you **use it**.

Always **obey** the law. State fishing regulations were created to protect the various species of fish found in your state. Make sure all your catches meet the state minimums for each species. Chances are you won't be caught if you don't, but that is not the point. The environment of the pond, lake, or reservoir you will be fishing is a very delicate balance. Your state wildlife agency has determined the size, season and limit for each species to **insure** their future availability.

A lot of anglers never keep anything they catch. They figure that the fish gave them a good fight and deserves to live for another day when some other fisherman can have the enjoyment of catching it again.

Always have a **use** for your catch. If the creel limit for largemouth bass in your state is 5, but you only have plans for 2, what is the purpose of keeping the other 3 fish, if your lucky enough to catch that many.

Water is the foundation of life for every living thing. People use alot of it every day and industry uses even more. Most people take **clean water** for granted. This is a real mistake. There is enough water for everybody, provided we keep it clean, and that is where the problem lies for the next generation.

Unfortunately, pollution is a major problem in many of the very rivers and lakes that supply our drinking water. Everyone has heard of **acid rain**. This is a real problem for many eastern lakes, reservoirs and rivers, but it is not the major problem.

In many communities, sewage contamination is the major problem. Untreated or partially treated sewage is being dumped into various water systems. This sewage contains bacteria that could cause serious **public health** problems.

Sewage contains **nutrients**. When these organisms die simultaneously, the resulting decomposition depletes oxygen in the water and kills fish.

Runoffs from farmlands are also causing similar problems, because the fertilizer in these runoffs contains large quanities of nutrients.

Siltation of lakes and reservoirs is another real problem. Silt carried from land developments and other sources into lakes and reservoirs reduces light penetration and thus, **photosynthesis**. As it falls to the bottom, it covers the delicate organisms living there and often destroys the eggs of the fish in the area.

You can do something about these problems. Always practice good ethical standards and keep up with local environmental legislation. Make sure your voice is heard. Also, have a healthy respect for these resources and make a personal commitment to help protect them.

GLOSSARY

ACTION	Describes how a rod bends. Rods that have most of the bend in the tip end are called fast action while rods that bend evenly throughout their length are labeled parabolic.
ADAPTATION	The characteristic of a species to accommodate its basic nature to its environment to satisfy its basic needs of reproduction, comfort (security), and food.
AERATOR	An electric air pump used to maintain oxygen levels in live wells or bait containers.
ALGAE	Simple, photosynthetic plants with uncellular organs of reproduction.
ANADROMOUS	Species include salmon, shad, perch, striped bass.
ANTIREVERSE LEVER	A lever or knob that prevents the reel handles from turning backwards as a fish tries to take line.
BACKLASH	Line tangled on a casting reel as a result of a cast made when the spool continues to revolve after the line has stopped going off the reel.
BAIL	A wire half-round device that spools the line onto an open face spinning reel.
BAIT-CASTING	Name given to casting equipment that uses a bait-casting reel. The reel is also called a "level-wind" reel.
BARBEL	A whisker-like projection from the jaws of some fish, such as carp or catfish. They help the fish smell and feel.
BLANK	The basic shaft of fiberglass or graphite or other rod material on which a rod is built.
CANE POLE	Equipment that employs a long, slender rod and no reel. Cane poles are some-

	times called "bank poles."
CLASS LINE	A fishing line which tests less than the nominal strength listed on the label. Used when a fish is to be entered into a certain IGFA record keeping category.
CONSERVATION	The wise use without waste of a natural resource
CREEL LIMIT	A term used by some fisheries agencies to indicate the number of fish, by species, that can be legally caught in one day.
DISSOLVED OXYGEN	(DO): The oxygen utilized by fish which is put into water by forces such as wind, plants, micro-organisms, etc.
DRAG SYSTEM	A system of soft and hard washers in any reel that serve as a braking mechanism to slow a fish as it takes line off of a reel. Drag pressure is usually set to 1/4 to 1/3 of the line test.
EPILIMNION	The warm layer of water above the thermocline.
ETHICAL	Judging behavior right or wrong based on a set of values or opinions.
EUTROPHIC	Lake classification or lake type used to describe bodies of water characterized by high levels of nutrients in proportion to their total volume of water.
FINGERLING	A juvenile fish, usually about several inches, or one finger length, thus the name.
FOOD CHAIN	Chain of organisms existing in any natural community through which energy is transferred. Each link in the chain feeds on and obtains energy from the one preceding it and it in turn is eaten by and provides energy for the one following it as the food chains in a community make up the food cycle or food web.
FRY	Small juvenile fish that have just hatched out of the egg and up to sev-

	eral inches long (at which point they become fingerlings).
GILL	An arch-like breathing organ located behind the gill cover on a fish's head.
HABITAT	The type of place where a species of fish lives
HOOK KEEPER	A small device, usually made of wire, to hold a lure or hook when the rod is not being used.
ICHTHYOLOGY	The branch of zoology that deals with fish, their classification, structure, habits, and life history.
IMPOUNDMENT	A natural or artificial place where water is collected and stored for use.
LAKE CLASSIFICATIONS	Broad categories of lake types; oligotrophic (infertile), mesotrophic (fertile), eutrophic (very fertile).
LATERAL LINE SYSTEM	System of sense organs present in aquatic vertebrates (fish) in pores or canals arranged in a line down each side of the body and in complicated pattern of lines on the head. It detects pressure changes including vibrations (low frequency sounds) in water.
OLFACTORY NERVES	Nerves that allow for the sense of smell.
OLIGOTROPHIC	Lake classification used to describe bodies of water characterized by low amounts of nutrients in proportion to their total volume water.
PFD	An abbreviation for Personal Flotation Device, the technical term for a life vest.
PHOTO-SYNTHESIS	In green plants synthesis of organic compounds from water and carbon dioxide using energy absorbed by chlorophyll from sunlight.
PREDATOR-PREY	A fish that feeds on other fish. An interdependence between a species and an accessible and suitable forage.

SCALE	1. A chitinous covering plate on a fish. 2. A method (scaling) of removing the scales from a fish for cooking and eating. 3. A gauge used to weigh fish.
SCHOOL (of fish)	A number of fish of a species that are grouped together for mutual advantage.
SEASON	The period of time during a year that a particular species of fish may be harvested.
SEDIMENTATION	The deposition or accumulation of sediment.
SKIRTED SPOOL	A type of spool found on open face spinning reels where a flange or extension extends from the rear of the spool to cover the cup and spool housing and prevent dirt and line from getting into the reel.
SNELLED HOOK	A hook that is pre-tied with a short length leader.
STILL FISHING	A method of fishing whereby the bait is cast to a likely spot and waiting for the fish to bite.
STRUCTURE	A term often used by anglers to designate any type of object or cover attractive to fish. Structure that fish relate too includes stumps, rock piles, log jams, piers, docks, boat houses, channel markers, points of land, etc.
SUSPENDED FISH	Fish which are hovering considerably above the bottom in open water.
THERMOCLINE	Temperature stratification in a body of water. Specifically, the layer of water where temperature changes at least one half a degree per foot of depth.
ULTRA-LIGHT	A name given to casting equipment that is reduced in size for casting small, lightweight lures.
WEEDLESS	A hook or lure that tends to pass through aquatic vegetation without picking any up.

QUESTIONS TO ASK

As you have probably noticed by now, fishing requires the practical application of a considerable amount of information. Hopefully, you have picked up a logical system that will give you the confidence to take on most any lake or reservoir in this country.

Make sure you have everything you will need before you go. Planning is very necessary. Good fishermen are always prepared.

As has been stated, local information is important, sometimes critical. Find the local tackle shop familiar with the lake you wish to fish and be sure to ask these questions.

1. Do you have a map of the lake? May I buy or study it? Can you show me the best locations?
2. What species are available and will I be better off with live bait or artificials? If artificials, what kind and color during the time I plan to fish?
3. What is the water temperature?
4. Where are the fish holding? Is the spawn complete? Are they in shallow or deep water?
5. Can I rent a boat? Where?
6. What is the size limit on these fish? How many can I keep?

REMEMBER, ALWAYS OBEY ALL SIZE MINIMUMS, THROW TRASH ONLY INTO A TRASH CAN OR BAG AND NEVER KEEP MORE FISH THAN YOU CAN USE.